Lecture Notes in Electrical Engineering

Volume 265

For further volumes:
http://www.springer.com/series/7818

Jan Haase
Editor

Models, Methods, and Tools for Complex Chip Design

Selected Contributions from FDL 2012

 Springer

Editor
Jan Haase
Institute of Computer Technology
Vienna University of Technology
Vienna, Austria

ISSN 1876-1100
ISBN 978-3-319-01417-3
DOI 10.1007/978-3-319-01418-0
Springer Cham Heidelberg New York Dordrecht London

ISSN 1876-1119 (electronic)
ISBN 978-3-319-01418-0 (eBook)

Library of Congress Control Number: 2013946029

© Springer International Publishing Switzerland 2014
This work is subject to copyright. All rights are reserved by the Publisher, whether the whole or part of the material is concerned, specifically the rights of translation, reprinting, reuse of illustrations, recitation, broadcasting, reproduction on microfilms or in any other physical way, and transmission or information storage and retrieval, electronic adaptation, computer software, or by similar or dissimilar methodology now known or hereafter developed. Exempted from this legal reservation are brief excerpts in connection with reviews or scholarly analysis or material supplied specifically for the purpose of being entered and executed on a computer system, for exclusive use by the purchaser of the work. Duplication of this publication or parts thereof is permitted only under the provisions of the Copyright Law of the Publisher's location, in its current version, and permission for use must always be obtained from Springer. Permissions for use may be obtained through RightsLink at the Copyright Clearance Center. Violations are liable to prosecution under the respective Copyright Law.
The use of general descriptive names, registered names, trademarks, service marks, etc. in this publication does not imply, even in the absence of a specific statement, that such names are exempt from the relevant protective laws and regulations and therefore free for general use.
While the advice and information in this book are believed to be true and accurate at the date of publication, neither the authors nor the editors nor the publisher can accept any legal responsibility for any errors or omissions that may be made. The publisher makes no warranty, express or implied, with respect to the material contained herein.

Printed on acid-free paper

Springer is part of Springer Science+Business Media (www.springer.com)

Preface

This book is the latest contribution to the LNEE series, and it consists of selected papers presented at the Forum on Specifications and Design Languages (FDL) 2012, which took place in September 2012 at Vienna University of Technology, Vienna, Austria.

FDL is a well-established international forum devoted to dissemination of research results, practical experiences, and new ideas in the application of specification, design, and verification languages to the design, modelling, and verification of integrated circuits, complex hardware/software embedded systems, and mixed-technology systems. Modelling and specification concepts push the development of new design and verification methodologies to system level, thus providing a means for model-driven design of complex information processing systems in a variety of application domains. One of the principal advantages of FDL is that it brings together several related thematic areas and gives an opportunity to gain up-to-date knowledge in many broad areas of the fast evolving field of system design and verification. In 2012, some additional key areas were covered in the form of special sessions and tutorials included in the conference program.

This book presents a collection of the best papers from FDL 2012 and covers the following topic areas:

- Assertion Based Design, Verification and Debug (ABD)
- Language-Based System Design (LBSD)
- Embedded Analog and Mixed-Signal Design (EAMS)
- UML and MDE for Embedded System Specification & Design (UMES)
- Special Sessions of FDL 2012

The papers were selected by the topic area program chairs Dominique Borrione (responsible for ABD), Martin Radetzki (responsible for LBSD), Christoph Grimm (responsible for EAMS), and Julio Medina (responsible for UMES).

The chapters of this book present recent and significant research results in the areas of design and specification languages for embedded systems, SoC, and integrated circuits. The objective of the book is to serve as a reference text for researchers and designers interested in the extension and improvement of the application of design and verification languages in the area of embedded systems.

I would like to take this opportunity to thank the members of the program committee who made a tremendous effort in revising and selecting the best papers for the conference and the most outstanding among them for this book. I would also like to thank all the authors for the extra work made in revising and improving their contributions to the book.

Finally, I would like to express my special thanks to Adam Morawiec and Jinnie Hinderscheit from ECSI, who made this book possible.

Vienna, Austria,   Jan Haase

Contents

1 Formal Plausibility Checks for Environment Constraints 1
Binghao Bao, Jörg Bormann, Markus Wedler,
Dominik Stoffel, and Wolfgang Kunz
  1.1 Introduction 2
  1.2 Circuit Models 3
  1.3 Properties of Circuits 3
  1.4 Environment Constraints 6
    1.4.1 Implementable Constraints 6
    1.4.2 Composability 8
  1.5 Plausibility Checks in Coverage Analysis for Property Sets 9
    1.5.1 Complete Interval Property Checking (C-IPC) 10
    1.5.2 Plausibility Checks 11
  1.6 Experimental Results 13
  1.7 Conclusions 15
  References 16

**2 Efficient Refinement Strategy Exploiting Component
Properties in a CEGAR Process** 17
Syed Hussein S. Alwi, Cécile Braunstein,
and Emmanuelle Encrenaz
  2.1 Introduction 17
  2.2 Our Framework 19
    2.2.1 Concrete System Definition 20
    2.2.2 Abstraction Definition 21
    2.2.3 Initial Abstraction 23
  2.3 Refinement 23
    2.3.1 Properties of Good Refinement 23
    2.3.2 Negation of the Counterexample 24
    2.3.3 Ordering of Properties 27
    2.3.4 Filtering Properties 29

2.4	Experimental Results	30
2.5	Negation of the Counterexample as a Complementary Strategy	33
2.6	Conclusion and Future Works	34
References		35

3 Formal Specification Level ... 37
Rolf Drechsler, Mathias Soeken, and Robert Wille
3.1	Introduction	37
3.2	Preliminaries	40
	3.2.1 Unified Modeling Language	40
	3.2.2 Natural Language Processing	41
3.3	Formal Specification Level	43
3.4	Mapping Natural Language Specifications to the Formal Specification Level	44
	3.4.1 Determine the Structure of the Design	44
	3.4.2 Determine the Properties of the Design	45
3.5	Checking Correctness at the Formal Specification Level	47
	3.5.1 Verification of Static Aspects	47
	3.5.2 Invariant Removal	48
	3.5.3 Verification of Dynamic Aspects	48
3.6	Mapping from Formal Specification Level to the Electronic System Level	49
3.7	Tool Support	49
3.8	Conclusion	50
References		51

4 TLM POWER3: Power Estimation Methodology for SystemC TLM 2.0 ... 53
David Greaves and Mehboob Yasin
4.1	Introduction	53
4.2	Our Approach: TLM POWER3	56
	4.2.1 Extended Generic Payload: Distance + Hamming	57
	4.2.2 Output Reports	61
4.3	Performance	62
4.4	Accuracy	65
4.5	Conclusion	67
References		67

5 SCandal: SystemC Analysis for Nondeterminism Anomalies ... 69
Jan Henrik Weinstock, Christoph Schumacher, Rainer Leupers, and Gerd Ascheid
5.1	Introduction	69
5.2	SystemC Simulation Concept	71
5.3	Related Work	72
5.4	Process Order Dependency Test	73
	5.4.1 Behavior Observation	74

	5.4.2	Detectable Anomalies	77
	5.4.3	Controlled Scheduling	78
	5.4.4	PEO Dependency Analysis	78
5.5		Experiments and Case Studies	82
	5.5.1	Synthetic Tests	83
	5.5.2	SoClib	85
	5.5.3	Parallel Simulation of Mixed-Level Multicore Platform	85
5.6		Conclusion and Outlook	86
References			87

6 A Design and Verification Methodology for Mixed-Signal Systems Using SystemC-AMS 89

Yao Li, Ramy Iskander, Farakh Javid, and Marie-Minerve Louërat

6.1		Introduction	89
6.2		Unified Platform Architecture	91
	6.2.1	SystemC AMS Extensions	92
	6.2.2	CHAMS Sizing and Biasing Engine	93
6.3		Proposed Levels of Abstraction	96
6.4		Implementation of the Unified Platform	97
	6.4.1	Comparator TDF Module	99
	6.4.2	end_of_elaboration() function	100
	6.4.3	initialize() function	101
	6.4.4	processing() function	101
6.5		Transient Analysis Method	102
6.6		Experimental Results	103
	6.6.1	Sizing and Biasing Procedure of the Two-Stage Comparator	103
	6.6.2	Simulation Results of a Two-Stage Pipeline ADC	105
6.7		Conclusion	106
References			106

7 Configurable Load Emulation Using FPGA and Power Amplifiers for Automotive Power ICs 109

Manuel Harrant, Thomas Nirmaier, Christoph Grimm, and Georg Pelz

7.1	Introduction	109
7.2	Related Work	110
7.3	First Experimental Setup	111
7.4	Load Modelling for Real-Time Evaluation	115
7.5	Evaluation of Lamp Model	119
7.6	Experimental Results	123
7.7	Conclusion and Outlook	125
References		126

8   Model Based Design of Distributed Embedded Cyber Physical Systems 127
Javier Moreno Molina, Markus Damm, Jan Haase, Edgar Holleis,
and Christoph Grimm
  8.1   Introduction 128
    8.1.1   Model-Based Design Approach 128
    8.1.2   Multi-domain Simulation 129
  8.2   Previous Work 130
  8.3   Methodology 131
    8.3.1   Requirements 132
    8.3.2   Functional Model 132
    8.3.3   Hardware/Software Co-design 133
    8.3.4   Deployment 133
  8.4   Models Implementation 133
    8.4.1   Functional Node Model 135
    8.4.2   Embedded Platform Model 136
    8.4.3   Propagation Model 137
    8.4.4   Network Protocol Stack 137
    8.4.5   Environment Interaction 138
  8.5   Simulating the Energy Management Application 139
  8.6   Conclusion and Future Work 142
  References 142

9   Model-Driven Methodology for the Development of Multi-level Executable Environments 145
Fernando Herrera, Pablo Penil, Hector Posadas, and Eugenio Villar
  9.1   Introduction 146
  9.2   Related and Previous Work 147
  9.3   Environment Modelling Methodology 149
    9.3.1   Environment Structure and Connection to the System 149
    9.3.2   Levels of Abstraction in the Specification
            of Environment Behaviour 157
    9.3.3   Modeling Several Scenarios 158
  9.4   Toolset 158
    9.4.1   SystemC Generation 159
    9.4.2   File Structure Generation 159
  9.5   SystemC Simulation with the System Performance Model 160
  9.6   Example 160
  9.7   Conclusions 162
  9.8   Future Work 162
  References 163

10   GREEN HOME: The Concept and Study of Grid Responsiveness ... 165
Slobodanka Tomic, Jan Haase, and Goran Lazendic
  10.1   Introduction 165
  10.2   Home Gateway Functions 168
  10.3   Demand Response 169

10.4	Grid Responsiveness Concept		170
	10.4.1	The Day-Ahead Exchange	171
	10.4.2	The Intra-day Exchange	172
	10.4.3	User Responsiveness	172
10.5	The Model of Home Activities		173
10.6	Forecasting of the Uncertainty Level		174
10.7	The Test Bed		174
10.8	Conclusions		177
References			178

11 Polynomial Metamodel-Based Fast Optimization of Nanoscale PLL Components 179

Saraju P. Mohanty and Elias Kougianos

11.1	Introduction		180
11.2	Proposed Novel Fast Analog/Mixed-Signal Design Flow		181
11.3	Related Prior Research		184
11.4	Design of PLL Component Circuits		184
	11.4.1	Phase Detector	185
	11.4.2	Loop Filter and Charge Pump	185
	11.4.3	LC Voltage Controlled Oscillator	185
	11.4.4	Frequency Divider	186
11.5	Proposed Approach for Generation of Fast and Layout-Accurate Metamodels		187
	11.5.1	Data Sampling	189
	11.5.2	Data Centering	190
	11.5.3	Stepwise Regression	190
	11.5.4	Verification of the Metamodel	190
11.6	Proposed Metamodel Based Design Optimization		192
11.7	Experimental Results		193
11.8	Summary, Conclusions, and Future Direction of Research		196
References			198

12 Methodology and Example-Driven Interconnect Synthesis for Designing Heterogeneous Coarse-Grain Reconfigurable Architectures 201

Johann Glaser and Clifford Wolf

12.1	Introduction		201
12.2	Development of Reconfigurable Hardware		202
	12.2.1	Pre-silicon Phase	203
	12.2.2	Post-silicon Phase	203
12.3	Design Methodology		203
	12.3.1	Specification	204
	12.3.2	Application Analysis	204
	12.3.3	Merge	206
	12.3.4	Implementation	206
	12.3.5	Verification	206

	12.3.6	Post-silicon Phase	207
	12.3.7	Tools	207
12.4		Interconnect for Reconfigurable Modules	207
	12.4.1	Common Topologies	207
	12.4.2	A Tree Topology	208
	12.4.3	Analysis of the Tree Topology	210
12.5		Interconnect Synthesis	211
	12.5.1	Optimization Algorithm	212
	12.5.2	Implementation Details	214
12.6		Evaluation of InterSynth	214
	12.6.1	Filter Networks	215
	12.6.2	Logic Networks	217
12.7		Yosys	218
12.8		Conclusion	218
References			220

Contributors

Syed Hussein S. Alwi Université Pierre et Marie Curie Paris 6, LIP6-SOC (CNRS UMR 7606), Paris, France

Gerd Ascheid Institute for Communication Technologies and Embedded Systems, RWTH Aachen University, Aachen, Germany

Binghao Bao University of Kaiserslautern, Kaiserslautern, Germany

Jörg Bormann University of Kaiserslautern, Germany

Cécile Braunstein Université Pierre et Marie Curie Paris 6, LIP6-SOC (CNRS UMR 7606), Paris, France

Markus Damm Technische Universität Kaiserslautern, Kaiserslautern, Germany

Rolf Drechsler Group of Computer Architecture, University of Bremen, Bremen, Germany

Cyber-Physical Systems, DFKI GmbH, Bremen, Germany

Emmanuelle Encrenaz Université Pierre et Marie Curie Paris 6, LIP6-SOC (CNRS UMR 7606), Paris, France

Johann Glaser Institute for Computer Technology, Vienna University of Technology, Vienna, Austria

David Greaves Computer Laboratory, University of Cambridge, Cambridge, UK

Christoph Grimm Technische Universität Kaiserslautern, Kaiserslautern, Germany

Jan Haase Vienna University of Technology, Institute of Computer Technology, Vienna, Austria

Manuel Harrant Infineon Technologies AG, Neubiberg, Germany

Fernando Herrera University of Cantabria, ETSIIT, Santander, Spain

Edgar Holleis Tridonic, Dornbirn, Austria

Ramy Iskander Université Pierre et Marie Curie, LIP6, Paris, France

Farakh Javid Université Pierre et Marie Curie, LIP6, Paris, France

Elias Kougianos Engineering Technology, University of North Texas, Denton, TX, USA

Wolfgang Kunz University of Kaiserslautern, Kaiserslautern, Germany

Goran Lazendic FTW Forschungszentrum Telekommunikation Wien GmbH, Austria

Rainer Leupers Institute for Communication Technologies and Embedded Systems, RWTH Aachen University, Aachen, Germany

Yao Li Université Pierre et Marie Curie, LIP6, Paris, France

Marie-Minerve Louërat Université Pierre et Marie Curie, LIP6, Paris, France

Saraju P. Mohanty Computer Science and Engineering, University of North Texas, Denton, TX, USA

Javier Moreno Molina Technische Universität Kaiserslautern, Kaiserslautern, Germany

Thomas Nirmaier Infineon Technologies AG, Neubiberg, Germany

Georg Pelz Infineon Technologies AG, Neubiberg, Germany

Pablo Penil University of Cantabria, ETSIIT, Santander, Spain

Hector Posadas University of Cantabria, ETSIIT, Santander, Spain

Christoph Schumacher Institute for Communication Technologies and Embedded Systems, RWTH Aachen University, Aachen, Germany

Mathias Soeken Group of Computer Architecture, University of Bremen, Bremen, Germany

Cyber-Physical Systems, DFKI GmbH, Bremen, Germany

Dominik Stoffel University of Kaiserslautern, Kaiserslautern, Germany

Slobodanka Tomic FTW Forschungszentrum Telekommunikation Wien GmbH, Vienna, Austria

Eugenio Villar University of Cantabria, ETSIIT, Santander, Spain

Markus Wedler University of Kaiserslautern, Kaiserslautern, Germany

Jan Henrik Weinstock Institute for Communication Technologies and Embedded Systems, RWTH Aachen University, Aachen, Germany

Robert Wille Group of Computer Architecture, University of Bremen, Bremen, Germany

Cyber-Physical Systems, DFKI GmbH, Bremen, Germany

Clifford Wolf Institute for Computer Technology, Vienna University of Technology, Vienna, Austria

Mehboob Yasin Computer Laboratory, King Faisal University, Al-Ahasa, Saudi Arabia

Chapter 1
Formal Plausibility Checks for Environment Constraints

Binghao Bao, Jörg Bormann, Markus Wedler, Dominik Stoffel, and Wolfgang Kunz

Abstract Functional verification of a System-On-Chip (SoC) module requires that the legal behavior of its environment is modeled as part of the verification IP. In early stages of the SoC design process so called environment constraints are used for this purpose. As long as a complete implementation of the environment is not yet available these constraints restrict the inputs of the device under verification to reasonable values.

Using such constraints during functional verification, however, imposes a high risk that legal environment behavior is pruned away. In this case some faulty behavior of the DUV may not be stimulated, i.e., the constraints may mask a bug.

Since the individual modules of an SoC are usually developed simultaneously it may not be possible to check the constraints against the environment of a module before integration. Detecting verification gaps due to overconstrained environment assumptions at this late stage of the design process, however, requires a step back into module verification and may compromise project closure.

In order to overcome this bottleneck of the verification flow we suggest two efficient plausibility checks for constraints that can be conducted without a concrete implementation of the considered environment. Our experimental results show that the proposed techniques detect issues that would otherwise remain undetected at least until module integration. The tests are applicable in both formal and constrained random verification environments.

B. Bao (✉) • M. Wedler • D. Stoffel • W. Kunz
University of Kaiserslautern, Kaiserslautern, Germany
e-mail: bao@eit.uni-kl.de; wedler@eit.uni-kl.de; stoffel@eit.uni-kl.de; kunz@eit.uni-kl.de

J. Bormann
University of Kaiserslautern, Germany
e-mail: Joerg.D.Bormann@web.de

J. Haase (ed.), *Models, Methods, and Tools for Complex Chip Design*, Lecture Notes in Electrical Engineering 265, DOI 10.1007/978-3-319-01418-0_1,
© Springer International Publishing Switzerland 2014

1.1 Introduction

Environment constraints are of major importance for the functional verification of System-On-Chip modules. Regardless of whether a design is verified using formal techniques or whether a classical simulation-based verification approach is chosen, a verification engineer in either case needs to model the environment of the device under verification (DUV) as well.

Classical directed testbenches constrain the behavior of the environment in a rather implicit manner by the set of stimuli generated during simulation. More advanced techniques such as constrained random simulation or formal assertion based verification require an explicit specification of environment constraints [10, 12].

The holy grail of formal verification is its promise to completely exercise the entire input, output and state space of a design in order to prove the absence of a bug. It has been demonstrated that formal property checking, if complemented with a coverage analysis for the property set [2,3,5,11] reaches this goal with reasonable verification effort. The coverage analysis identifies specification gaps in cases where particular input scenarios are not covered by any of the specified properties. Again environment constraints are used to restrict the analysis to relevant scenarios.

Verifying a design with respect to a constrained environment always bears the risk of masking bugs. Moreover, such overconstraining is more difficult to detect than other specification mistakes as it does not show up in a counterexample for a failing property. In industrial practice the review of constraints is thus taken very seriously. However, sometimes a collection of environment constraints may imply subtle consequences that may not be immediately obvious to the verification engineer. In particular, it may be the case that a collection of fairly simple constraints, where each individual constraint has a very reasonable intention, turns out to be problematic when the constraints are applied in their entirety.

Vacuity checking may guarantee that at least one input scenario exists such that the constraint is satisfied. It is a standard technique in today's verification tools to approach overconstraining. However, it may only identify the case where a constraint cannot be satisfied at all and leaves many other constraint issues undetected.

In this paper we explore additional plausibility tests for environment constraints in a SAT-based property checking environment [1, 6–8]. We demonstrate that these checks can identify critical constraint issues and may guide the user to specify consistent constraints. We also envision that the presented techniques may likewise be useful in a constrained random simulation environment as well.

In particular we introduce the notion of implementability and loop-free composability of constraints. This formalizes the reasonable requirements that there should exist at least one environment that is able to fulfill the constraints in such a way that the composition of environment and design is a valid circuit, i.e., the composition does not contain any combinational loops in its logic.

We demonstrate the effectiveness of our techniques with a case study conducted in an industrial setting. We analyze constraints for a verification IP specifying the protocol compliance of an Infineon device for the Flexible Peripheral Interface

1 Formal Plausibility Checks for Environment Constraints

(FPI) bus. The formal property checker OneSpin 360 DV [8] is used to check the properties against the device and to prove the completeness of the verification IP. The entire verification is conducted under the assumption of a set of environment constraints. These constraints describe, for example, protocol-compliant behavior of the bus signals that are inputs to the DUV.

The remainder of the paper is organized as follows: Sects. 1.2 and 1.3 introduce basic formalisms for specifying circuit models and properties. Environment constraints are introduced in Sect. 1.4. Also, this section introduces checks for implementability and composability of constraints. These checks are evaluated in our experiments in Sect. 1.6 before the paper concludes with Sect. 1.7. Furthermore in Sect. 1.5 we also explore the usability of these checks in coverage analysis for property sets.

1.2 Circuit Models

In this paper we model the behavior of sequential circuits by binary-encoded finite state machines and corresponding Kripke models. A finite state machines (FSM) $M = (I, O, S, S_0, \delta, \lambda)$ consists of a set of input values I, a set of output values O, a state set S with a subset $S_0 \subset S$ of initial states, and two functions $\delta : I \times S \to S$ and $\lambda : I \times S \to O$ called state transition function and output function, respectively. An FSM M with $I \subset \mathbb{B}^n, O \subset \mathbb{B}^m, S \subset \mathbb{B}^k$ and $\mathbb{B} = \{0, 1\}$ is called binary-encoded. In this case δ and λ are multi-output Boolean functions and may be represented by a Boolean network.

Similarly, a Kripke model $K = (S, S_0, T, A, L)$ is a finite state transition structure with a state set S, initial state set $S_0 \subset S$, a state transition relation $T \subset S \times S$, a set of atomic formulas A and a labeling function $L : A \to 2^S$.

An FSM can be converted into a corresponding Kripke model where the information about inputs and outputs becomes part of the state information, i.e., the states of the Kripke model are triples $\hat{s} = (i, o, s)$. For better readability, we also simply use the symbol s to denote the state of a Kripke model if we do not care about its individual components. In the sequel we only consider Kripke models that are derived from a corresponding FSM. Sometimes we are interested in specifying or observing the internal signals of the Boolean networks representing the transition or output function of a binary-encoded FSM. In this case, we may assume the states of the corresponding Kripke model to be labeled accordingly. We use both models interchangeably depending on which is more convenient in a particular context.

1.3 Properties of Circuits

The front-end of a property checker computes circuit models from the description of a design given in some hardware description language. Both, the model and the design, describe the behavior of a DUV in its entirety.

By contrast, properties focus on a particular aspect of the design behavior and describe the corresponding behavior in a concise way. The restriction of the individual property to a particular scenario results in more compact formal specifications that are also easier to comprehend and maintain. In this paper we use a property checking technique called interval property checking [11]. This variant of SAT-based property checking conducts unbounded proofs for safety properties formulated in terms of design signals within a bounded time interval. Conceptually, interval properties can be viewed as LTL safety properties Gp where p is a formula that combines atomic formulas using only the Boolean connectives and the operator X.

The formula p of such an interval property characterizes a set of finite state sequences. By contrast the safety property Gp characterizes infinite state sequences, also referred to as *traces*.

We use state predicates $\eta(s)$ to characterize state sets and state sequence predicates $\sigma_l(s_0, \ldots, s_l)$ of length l to characterize sets of state sequences $\pi_l = (s_0, \ldots, s_l)$. If the length l of a sequence predicate matters we also call it an l-sequence predicate. State predicates can be considered as 0-sequence predicates. Both types of predicates can be defined using the input, state and output variables of the FSM or the corresponding state variables of the Kripke model. If only input variables of the FSM are used we call the corresponding predicate an *input predicate* or *input trigger*. Similarly, output predicates are defined using only output variables, and state predicates only depend on state variables (of the FSM). Predicates that only evaluate the value of a single state variable v, input variable i or output variable o are called *elementary state predicates* and are denoted by $v(s)$, $i(s)$ and $o(s)$, respectively. We allow every l-sequence predicate π_l to be applied to m-sequences $\pi_m = (s_0, \ldots, s_m)$ with $m \geq l$. In this case the predicate is evaluated on the l-prefix $\pi_l := (s_0, \ldots, s_l)$ of π_m, and the tail sequence $(s_{l+1}, \ldots, s_m)$ remains unrestricted. This guarantees that the usual Boolean operators $\vee$, $\wedge$, $\neg$ and $\Rightarrow$ are also applicable to sequence predicates that may possibly have different lengths. The maximum length l_{max} of the operands to these operators then determines the length of the resulting predicate. Sequence predicates can also be shifted in time using the *next* operator with

$$next(\sigma_l, n)\big((s_0, s_1, \ldots, s_{n-1}, s_n, s_{n+1}, \ldots, s_{n+l})\big)$$
$$:= \sigma_l\big((s_n, s_{n+1}, \ldots, s_{n+l})\big).$$

Using this operator we can define a concatenation operation $\odot$ for l-sequence predicates:

$$\sigma_l \odot \sigma_k := \sigma_l \wedge next(\sigma_k, l)$$

The predicate $\sigma_l \odot \sigma_k$ characterizes $k + l$-sequences $\pi_{k+l} = (s_0, \ldots, s_l, \ldots, s_{k+l})$ where s_l is evaluated as ending state of σ_l and starting state of σ_k. Non-overlapping

1 Formal Plausibility Checks for Environment Constraints

concatenation can be expressed using the special l-sequence predicate $any_l(\pi_l)$ that evaluates to true for every sequence π_l, as follows:

$$\sigma_l \oplus \sigma_k := \sigma_l \odot any_1 \odot \sigma_k$$

To conclude the introduction of sequence predicates it should be noted that every sequence predicate can be defined using elementary state predicates, the *next* operator and the Boolean connectives defined on sequence predicates. It is even sufficient to apply the *next* operator only directly to the elementary state predicates. Such a representation of a sequence predicate σ_l is called a *timed normal form*.

If v and v' are state variables we write $\sigma_l[v \leftarrow v']$ for the sequence predicate that is derived from σ_l by substitution of every occurrence of v by v'. If substitution is only performed for a particular state s_k of the sequence we write $\sigma_l[v(s_k) \leftarrow v'(s_k)]$. Substitution with constants yields the co-factors $\sigma_l(\pi_l)|_{v(s_k)} := \sigma_l[v(s_k) \leftarrow 1]$ and $\sigma_l(\pi_l)|_{\neg v(s_k)} := \sigma_l[v(s_k) \leftarrow 0]$. This allows us to quantify out particular state variables v at a particular timepoint $k \in \{0 \ldots l\}$ from a sequence predicate σ_l. We write $\forall_{v(s_k)} : \sigma_l$ to denote the sequence predicate

$$(\forall_{v(s_k)} : \sigma_l)(\pi_l) := \sigma_l(\pi_l)|_{v(s_k)} \wedge \sigma_l(\pi_l)|_{\neg v(s_k)}.$$

As a short notation for quantifying out the state variable v at every timepoint $k \in \{0 \ldots l\}$ we write

$$(\forall_{v(\pi_l)} : \sigma_l)(\pi_l) := (\forall_{v(s_0)} \ldots \forall_{v(s_l)} : \sigma_l)(\pi_l).$$

Similarly, the existential quantifiers $\exists_{v(s_k)} : \sigma_l$ and $\exists_{v(\pi_l)} : \sigma_l$ can be defined.

The characteristic function of the transition relation $T(s_0, s_1)$ of a Kripke model is an important 1-sequence predicate for property checking. It can be used to determine state sequences that correspond to valid paths in the Kripke model. Such paths are characterized by the l-sequence predicate *ispath*:

$$ispath_l(\pi_l) := \left(\bigodot_{i=1}^{l} T\right)(\pi_l) = \bigwedge_{i=1}^{l} T(s_{i-1}, s_i).$$

Note that the starting state s_0 of an l-sequence satisfying the *ispath* predicate is not restricted to initial states of the Kripke model. Therefore, *ispath* is useful for proving the unbounded validity of the safety property

$$G\sigma_l := \bigwedge_{t \geq 0} next(\sigma_l, t)$$

specified by σ_l. If $ispath_l \Rightarrow \sigma_l$ is a tautology then $G\sigma_l$ is valid unboundedly. This check for tautology can be conducted effectively using a satisfiability (SAT) solver. For simplifying notations we sometimes omit the operator G if it is clear from the context that the corresponding safety property of σ_l is considered.

1.4 Environment Constraints

So far we used sequence predicates only to describe the intended behavior of a circuit. In practice, circuits are rarely designed to work in any arbitrary environment. The communication of a circuit with its environment is usually restricted by some sort of protocol. In order for a circuit to show its intended behavior we need to assume that the environment complies with this protocol. For modeling legal environment behavior by environment constraints we also use sequence predicates. Such constraints need to be selected carefully in order to ensure that they can actually be fulfilled by the environment and that they do not overconstrain the design.

This is of particular interest when different portions of a design are developed by different IP providers and the final environment for the design is not known prior to the integration phase. At that stage of the project it is often too late if inconsistent environment constraints for the individual modules are detected and the required changes may cause a project to fail or miss its deadline.

Here we propose two plausibility checks for the environment constraints of a module that can be checked without a concrete environment at hand.

1.4.1 Implementable Constraints

It is a reasonable requirement for a sequence predicate σ_l that is to be used as environment constraint for a circuit model M that at least one environment must exist that can fulfill the constraint. In other words, we need to ensure that an FSM M_E (modelling the environment) exists that satisfies σ_l. This environment model computes the inputs to the DUV using the DUV's outputs and possibly also internal signals from the DUV that are used to specify σ_l. In addition it may use additional free inputs to model non-determinism in the constraint. If such a model M_E exists we call the constraint σ_l *implementable*. For implementable constraints one may generate a *most-general* implementation that can exhibit every behavior that is not explicitly forbidden by the constraint. In the literature such an implementation is also known as *can-do object* [9]. Another way to derive such an implementation is to apply the synthesis techniques of [4].

In our context we are only interested in the *existence* of an environment constraint's implementation. For typical constraints encountered in practice this question can be decided without a full-blown synthesis of can-do objects. In the sequel we will develop such a test.

As an example of a non-implementable environment constraint consider the sequence predicate $i = next(o)$ for an input i and an output o of M. For the environment M_E, i is an output and o is an input. Unless the output is constant, the constraint models a precognitive, i.e., non-causal, environment that obviously does not exist.

1 Formal Plausibility Checks for Environment Constraints

Checking that an arbitrary sequence predicate is implementable is a non-trivial task. To simplify this task we restrict the syntax of the language used to specify the environment constraints σ_l such that only causal constraints can be formulated. An input i may be restricted by a *basic constraint* of the form

$$c^i(\pi_l) := i(s_l) \triangleleft \sigma_l^i(\pi_l), \quad \triangleleft \in \{\Rightarrow, \Leftarrow, =\}.$$

This guarantees that the input i does not depend on future values of a signal trace. Moreover, we assume that only input and output variables are used to express the basic constraints c^i. In the sequel we only consider constraints σ_l that are specified as Boolean expressions over the basic constraints c^i.

The individual constraints c^i are, obviously, implementable and thus causal. However, this does not guarantee that σ_l is also implementable. To see this, consider the state predicate $i \Leftarrow o_1 \wedge i \Rightarrow \neg o_2$ which places two constraints on the input i depending on two current outputs o_1, o_2 of the module. This restricts o_1, o_2 such that always one of these outputs has to take the value 1. However, the outputs of the module are inputs to its environment and cannot be controlled by any circuit implementing the environment. Thus, the constraint is not implementable.

In order to guarantee implementability we check whether the constraint σ_l can be fulfilled regardless of the output sequence generated by the module and regardless of the previous inputs that have been asserted. This yields the following QBF formula in terms of the output variables $o_1, \ldots, o_n$ and the inputs variables $i_1, \ldots, i_m$ used in the timed normal form of σ_l:

$$\forall_{o_j(\pi_l), j=1,\ldots,n} \forall_{i_h(\pi_{l-1}), h=1,\ldots,m} \exists_{i_h(s_l), h=1,\ldots m} :$$

$$\sigma_l(s_0, \ldots, s_l)$$

The constraint σ_l is implementable if the above QBF formula is a tautology. The intuition behind this formula is the following. In order for the environment constraint σ_l to be implementable it is necessary that the outputs of the environment (i.e., the outputs of the DUV) are in a causal functional relationship with the inputs of the environment (i.e., the outputs of the DUV). This is the case if for every combination of DUV output l-sequence and DUV input $l - 1$-sequence ("history") there exists a ("current") DUV input complying with the constraint. (No future time references are allowed in the constraint.) A construction of a circuit corresponding to and implementing the constraint can be found in [2]. To disprove implementability it is thus sufficient to check the following formula for satisfiability:

$$\exists_{o_j(\pi_l), j=1,\ldots,n} \exists_{i_h(\pi_{l-1}), h=1,\ldots,m} \forall_{i_h(s_l), h=1,\ldots m} :$$

$$\neg \sigma_l(s_0, \ldots, s_l)$$

We solve this formula by explicit elimination of the inner universal quantifier and a propositional SAT solver. In our application this is feasible because most often σ_l is merely a conjunction of basic constraints where an input in state s_l only depends on a very limited number of other inputs. The negation, $\neg \sigma_l$, is thus a disjunction of the negated basic constraints. We group the inputs with respect to the dependency relation induced by the basic constraints such that no dependency between inputs of different groups exists. In this case we may individually quantify the inputs for each group and consider their disjunction as final formula to be treated by the SAT solver.

1.4.2 Composability

We model a design under verification and its environment based on finite state machines. Properties describe the behavior of a system only at discrete synchronous time points (usually with reference to a clock signal) and in steady-state conditions. A property checker for this kind of properties cannot verify asynchronous behavior that results from combinational loops. In order to ensure that our environment constraints only model environments whose implementation is compatible with this computational model and verification technique we need to exclude combinational loops between the environment and the DUV.

Therefore, the second plausibility test for environment constraints that we suggest considers the composition of the module with a hypothetical implementation of the constraint and makes sure that no combinational loops are created by this composition. (It is obviously only applicable to implementable constraints.)

At a first glance, it seems appealing to require that *every* implementation of an environment constraint could be safely composed with our module. However, it turns out that this requirement is too strict. Consider a module with inputs i_1, i_2 and an output o. Assume a combinational dependency between o and i_2 but not between o and i_1. Consider the environment constraint $i_1 = o$. Every environment connecting i_1 with o implements the constraint. Under all such environments only those are composable that do not connect i_2 and o. Obviously this is not an issue of the constraint that is not talking about i_2 at all.

Instead of considering all implementations of the constraint we may resort to an existential requirement. In this case we would require that at least one environment for the module exists such that it implements the constraint and does not produce any combinational loops in the composition with the module M. However, this may cause problems as well. To see this consider the constraint $x_1 = y$ for a 2-input AND gate with inputs x_1, x_2 and output y. In this case the constraint can be implemented by connecting the output y with the input x_1. This leads to a combinational loop in the composition. Certainly this implementation in some sense is the most reasonable implementation of the constraint and we would like to forbid this type of constraint to prevent the resulting combinational loops. However, there are alternative implementations. For example the circuit producing a constant zero output for x_1 also turns out to be a valid implementation of the constraint. Additionally, this implementation does not produce a combinational loop with the AND gate.

The requirement that only a single implementation of a constraint be composable with the module M without combinational loops is, therefore, too weak. To rule out over-constraining implementations like the constant-zero implementation for the above mentioned constraint we take a look at *most-general* implementations. A most-general implementation of a constraint, also referred to as can-do object [9], is a circuit that can produce every behavior that is not explicitly forbidden by the constraint. Note that neither the constant-zero implementation of the previous example nor the implementation connecting o and i_2 in the earlier example are such most-general implementations because they restrict behavior that is definitely allowed by the respective constraints.

It turns out that most-general implementations are well suited for the definition of loop-free composable constraints:

Definition 1. An implementable constraint α is called loop-free composable with a module M if a most-general implementation M_α of α exists that can be composed with M without introducing combinational loops.

A naive way of checking loop-free composability as defined above would explicitly generate the most-general implementation. (Note that this would be possible with the synthesis techniques of [9].) However, as we are interested in combinational dependencies rather than exact functionality of these implementations we may resort to a much simpler structural analysis conducted on the original constraints. In this analysis we again benefit from the restriction of our constraint language. We simply extract the combinational dependencies from the basic constraints $c^i(\pi_l)$ used in our framework. Every pair of inputs and outputs that occurs with the same temporal offset of the *next* operator is considered as a potential combinational dependency. (Due to the causality of the basic constraints there cannot be any other such dependencies.) The dependencies derived from the constraint in this way are added to the signal dependency graph for the inputs and outputs of the circuit model M of the DUV. The resulting dependency graph for the system is then analyzed for loops. Note that in cases where the model of the DUV is not yet available because formal specification and design are developed in parallel one may refer to the properties of the design and conduct a similar analysis as for the constraint to generate a signal dependency graph. In either case an acyclic extended dependency graph guarantees the existence of a safe environment.

1.5 Plausibility Checks in Coverage Analysis for Property Sets

The quality of property sets in formal property checking can be measured by different coverage metrics. Some of these coverage metrics assess whether the functional behavior of the design under verification is fully captured by the property set. Usually the coverage analysis is conducted on the individual modules rather than the entire SoC design either due to the computational complexity, or due to the limitation of applied design/verification methodology. On account of this, reactive environment constraints are also needed for the coverage analysis for property sets in order to restrict the analysis to relevant scenarios.

In this section we explore the usability of the plausibility checks presented in Sect. 1.4 in a setting where Complete Interval Property Checking (C-IPC) [2, 11] is used as basic property checking approach for SoC modules. In this approach, a set of safety properties is called a *complete specification* or simply *complete* if it *uniquely* describes the design [2, 3]. Although our approach is presented in the context of C-IPC, it can be used also with other coverage analysis methods applying environment constraints.

1.5.1 Complete Interval Property Checking (C-IPC)

Complete Interval Property Checking is a SAT-based property checking technique [2, 8, 11] that guarantees that a given property set $P = \{p_1, \ldots, p_n\}$ uniquely determines the output behavior of a particular device under verification. The properties p_i describe the temporal behavior of the signals $s_1, s_2, \ldots, s_j$ of a device. In order to analyze completeness of a property set independently of a concrete implementation we can map the properties onto a second set of fresh signal variables $s'_1, \ldots, s'_j$ and denote this by $P' = \{p'_1, \ldots, p'_n\}$.

In addition, a module also carries so called determination conditions that specify under which circumstances a particular input or output has to be uniquely determined. For example, a data signal on a bus may be guarded by a valid flag and only needs to be determined if this flag is asserted. We consider determination conditions for both inputs and outputs and refer to them as determination assumptions and determination requirements of the DUV, respectively. A determination condition d_s for a signal s is a property that evaluates to true whenever the signal s needs to be determined.

In particular, we assume determination assumptions d_i and determination requirements d_o to be specified for each input i and each output o. By default we assume $d_i = d_o = \textit{true}$ for all ports i, o of the device, i.e., we assume all inputs to be determined at every point in time and require the same for the outputs. However, the verification engineer may overwrite these default values by weaker conditions. For checking the validity of a determination condition d_s we will use two independent copies s and s' for each signal used within d_s. The property $D_s := (d_s \vee d_{s'}) \Rightarrow (s = s')$ indicates whether a signal s is determined. With this notation we can precisely define how the determination conditions should be interpreted:

Definition 2. A property p completely specifies a module M with respect to determination assumptions $d_{i_1}, \ldots, d_{i_n}$ and determination requirements $d_{o_1}, \ldots, d_{o_m}$ iff the following property is a tautology:

$$\bigwedge_{j=1}^{n} D_{i_j} \wedge p \wedge p' \Rightarrow \bigwedge_{j=1}^{m} D_{o_j}. \tag{1.1}$$

1 Formal Plausibility Checks for Environment Constraints 11

If $p = \bigwedge_{j=1}^{k} p_j$ completely specifies a module M then the property set $P = \{p_1, \ldots, p_k\}$ is called a complete property set with respect to the determination conditions d_{i_j}, d_{o_j}.

Note that for the default determination conditions $d_i = d_o = true$ this definition states that any two modules satisfying all the properties need to be sequentially equivalent.

In practice, verification of an SoC module needs to consider so called environment constraints that model the behavior of a realistic environment for a DUV. In this work, we model such restrictions by reactive constraints that evaluate previous and current outputs of the device to determine ranges of valid values for the inputs to the device.

By replacing Eq. 1.1 in Definition 2 with Eq. 1.2 we obtain a criterion for checking completeness of a module with respect to its (reactive) environment constraints.

$$c \wedge c' \wedge \bigwedge_{j=1}^{n} D_{i_j} \wedge p \wedge p' \Rightarrow \bigwedge_{j=1}^{m} D_{o_j} \tag{1.2}$$

Note that c in this equation denotes the conjunction of all environment constraints.

1.5.2 Plausibility Checks

When every module is verified completely with respect to the default determination conditions and no constraints are applied, the completeness results for the modules guarantee sequential equivalence of alternative implementations. However, as soon as we add environment constraints to the completeness analysis, things become more complicated. It is obvious that the completeness results for the property set are not trustable anymore when we apply environment constraints that violate plausibility checks presented in Sect. 1.4 to the completeness analysis. Recall that in this case the potential overconstraining may mask bugs in the design as well as verification gaps in property sets.

Yet for all that, using a constraint passing the plausibility checks cannot guarantee that the completeness results are valid. We illustrate this by means of an example.

Consider a D-flipflop M with a data input x and a data output y. We assume that M is used in an environment where the output y is connected with the input x via an inverter. To verify the D-flipflop with regard to this environment we may use an environment constraint $c := (x = \neg y)$. With this constraint the output y of M is uniquely determined even for the property $p :- G(true)$ that does not restrict the behavior of D-flipflop at all. The constraint $c := (x = \neg y)$ is, obviously, implementable, e.g., by the inverter. Thanks to the D-flipflop, it is also loop-free composable with M. However, the property set $\{p\}$ is not a complete description of the D-flipflop's behavior.

In the context of completeness analysis we extend the implementability check to consider the determination assumptions together with the environment constraints. We will see that this is an important aspect. For ease of explanation we group the determination assumptions with environment constraints into an *assumption of completeness*.

As the first step let us reconsider the example above. The constraint c is implementable, however, together with the implicitly taken determination assumption $d_x = \textit{true}$ of module M this constraint already determines the output y of the module under verification M, i.e., it holds

$$c' \wedge c \wedge (x = x') \Rightarrow (y = y').$$

Every valid implementation of this assumption of completeness thus has to ensure that its input y is determined. This is obviously impossible as a circuit does not have control over its inputs. The determination of a module's outputs like y is always in the responsibility of the module itself. If the module does not determine this output then the assumption of completeness does not determine it either.

The test for implementability of constraints as developed in Sect. 1.4.1 solves a QBF satisfiability problem that ensures the existence of a current input for every history of the inputs and every history of the output including the current value. The assumption of completeness can be handled in the same manner except that two versions of signals need to be considered. In the assumption of completeness of our example this would require that for every value of y and every value of y' corresponding values of x and x' need to be found such that the assumption of completeness

$$c(x, y) \wedge c'(x', y') \wedge x = x'$$

is fulfilled. For $y \neq y'$ this is obviously unsatisfiable. Therefore, the assumption of completeness for the flipflop is not implementable. A reasonable way to resolve this issue would be to remove the assumption that the input is determined, i.e., to allow for using the flipflop in an undetermined environment. In this case the assumption of completeness would consist of the two versions of only the constraint c and obviously would be implementable. Note that under this corrected assumption of completeness the completeness checker would detect that the trivial property p leaves the output undetermined.

For the same reason introduced at the beginning of Sect. 1.4.2, we also need to exclude the combinational loops induced by the assumption of completeness during completeness checking. As we only consider implementable assumptions of completeness, we may resort to a most-general implementation of the conditions and take the combinational dependencies from there. To analyze combinational loops, we may take the method presented in Sect. 1.4.2. However, unlike in the analysis of a pure constraint, the method is here performed on two sets of variables from two modules M and M' (copy of M), respectively.

1.6 Experimental Results

The plausibility checks for environment constraints developed in this work have been successfully evaluated in an industrial setting. As a case study we consider a formal specification of a master interface for Infineon's Flexible Peripheral Interconnect (FPI) bus. The specification ensures compliance of a particular implementation with the FPI bus protocol. This protocol is quite similar to the industry's standard AMBA bus protocol. It includes features like pipelining of transactions to improve the throughput of the bus.

The main task of the considered master interface is to adapt a particular processor interface for reading and writing information from/to peripherals to the particular FPI bus protocol. Each interface transaction of the processor is mapped onto a respective protocol transaction on the bus. Due to the pipelining features of the protocol the interface may handle two concurrent requests from the processor.

An industrial design of such an FPI bus master interface has been completely verified by applying the OneSpin 360 DV interval property checker with its extended features for completeness checking. Completeness of the specification was derived under an environment constraint composed of 72 basic constraints involving 36 signals.

The inputs and outputs of the considered design can be grouped into two categories. The first group is used in the communication with the processor while the other group forms the actual interface to the bus.

After thorough review of the environment constraint for the DUV by the verification and design teams the verification engineer was convinced that the constraint precisely captures the required environment behavior under which the DUV was supposed to meet its specification.

However, the plausibility checks developed in this work revealed two serious issues that may easily have masked a bug if they had remained undetected. For the check the dependency relation between signals referred to by the constraint was used to decompose it into nine groups of basic constraints referring to pairwise disjoint sets of signals. The plausibility checks were individually applied to each group. In one of the groups two bugs were detected. This group still has a source code size of about 30 lines.

This explains why under these circumstances the subtle interdependencies detected by the plausibility checks have been overlooked by the verification team. By contrast, our fully automatic plausibility checks analyzed each group within 190 ms using less than 95 MB of memory on an Intel Core i7 CPU 860 at 2.8 GHz with 8 GB of RAM.

In the case of the above-mentioned issues the code fragments that caused the plausibility tests to fail stem from distant locations within the constraint specification. This is true even though the verification engineer put significant effort into ordering the basic constraints in such a way that expressions referring to similar sets of signals are located close to each other.

```
if bus_is_idle_i then
        next(ready_i) = '1';
end if and
if next(this_master_is_driving_bus_o) then
        next(ready_i) = next(ready_o);
end if;
```

Fig. 1.1 Pseudo-code related to the first bug

```
if next(this_master_is_driving_bus_o) then
        next(ready_i) = next(ready_o);
else if bus_is_idle_i then
        next(ready_i) = '1';
end if;
```

Fig. 1.2 Possible solution to the first bug

In the sequel we use a pseudo-code notation to illustrate the nature of the identified issues within our constraints. For reasons of space we only present the relevant subexpressions that actually form the bug.

The first issue identified by the plausibility checks is related to the $ready\_i$ signal in the FPI bus. Figure 1.1 illustrates two subexpressions of the corresponding constraint that introduce an issue regarding implementability.

Within these code snippets the suffix 'i' indicates that a signal is an input of the DUV, whereas the suffix 'o' identifies outputs.

The code fragment states that whenever the FPI bus is in the idle state, at the next time point the $ready$ signal of the bus should have the Boolean value '1'. The latter indicates that there must be a master or a slave in the system selected to drive the bus signals. Furthermore, the second part of the code fragment considers the case that the DUV itself is selected to drive the bus. In this case the $ready\_i$ signal should actually correspond to the value of $ready\_o$ provided by the DUV.

In turns out that this constraint fails within the implementability check. A counterexample shows that for the present time point, whenever the previous value of $bus\_is\_idle\_i$ is logic '1', and the DUV is driving the bus, and the value of $ready\_o$ is logic '0', no value for the signal $ready\_i$ is available to satisfy the constraint. Actually, this constraint has the subtle consequence that if at a previous time point the bus is in the idle state and at present the DUV is selected to drive the bus, the $ready\_o$ must be '1'.

Note that this implication of the constraint is actually one of the verification goals for the DUV, i.e., the constraint would have masked cases where this requirement is not fulfilled. To resolve the problem we use a cascaded if-then-else structure to formulate the constraint as listed in Fig. 1.2.

The second issue of our constraints was discovered by the loop-free composability check specified in Sect. 1.4.2. The pseudo code in Fig. 1.3 illustrates a constraint

1 Formal Plausibility Checks for Environment Constraints

```
if lock_req_o = '1' and active_o = '1' then
        grant_i = '1';
end if;
```

Fig. 1.3 Pseudo code related to the second bug

to the input *grant_i*. It states that whenever the DUV is active and locks the request line the arbiter should grant this request immediately. The output *active_o* is not a part of the protocol. With this signal the DUV may indicate whether it still needs the bus or whether the bus could be put into a power saving mode. It should be connected to a power management unit in the environment of DUV. This unit may then determine whether the bus is in use by some other system components and otherwise assert additional signals to set the bus in the sleep mode as well. Obviously, the above constraint introduces a combinational dependency between *active_o* and *grant_i*. Unfortunately, these two signals are also in a combinational relationship in the circuit model of the DUV. Hence, the constraint causes a combinational loop with the DUV. This combinational loop may overconstrain the design because the verification tool can only consider the steady-state behavior caused by this loop.

Further investigation of the constraint reveals that the signal *active_o* is not necessary to formulate the constraint. The request line of the design can only be locked if the design is active at the same time. This relationship between the outputs should be checked by the verification IP of the module and not be implied from the environment constraint. Removing the *active_o* signal from the constraint is therefore a possible solution in this case.

1.7 Conclusions

In this paper we presented a set of formal plausibility tests for environment constraints of a verification IP. The tests can be applied in early stages of the design and verification process where a complete model of the environment may not yet be available. In current design and verification flows the issues identified with our tests are usually detected in the integration phase and then require backtracking to the module verification phase. We envision our technique to remove a bottleneck in the verification and design process.

We check automatically whether a constraint is implementable at all and whether a general implementation may introduce combinational loops with the design. This identifies cases where a constraint overconstrains the DUV. Obviously any constraint violating this requirement is erroneous.

Even for a constraint of moderate size with a few dozens up to even a few hundreds of lines of code detecting flaws of this kind by manual code review is tedious and error-prone and an automatic technique as proposed in this work is highly desirable.

Our experimental results indicate that the tests are very efficient in terms of computational complexity. The tests revealed two serious issues in environment constraint of an industrial verification IP. If undetected such issues could easily mask severe design bugs.

The technique is applicable in both formal and constrained random simulation-based verification environments.

References

1. Biere, A., Cimatti, A., Clarke, E.M., Fujita, M., Zhu, Y.: Symbolic model checking using SAT procedures instead of BDDs. In: Proceedings of the International Design Automation Conference (DAC), New Orleans, LA, USA, pp. 317–320 (1999)
2. Bormann, J.: Vollständige Verifikation. Dissertation, Technische Universität Kaiserslautern (2009)
3. Claessen, K.: A coverage analysis for safety property lists. In: Proceedings of the International Conference on Formal Methods in Computer-Aided Design (FMCAD), pp. 139–145. IEEE Computer Society, Austin, Texas, USA (2007)
4. Langer, J., Heinkel, U.: High level synthesis using operation properties. In: Proceedings of Forum on Specification Design Languages (FDL 2009), Sophia Antipolis, France, pp. 1–6 (2009)
5. Lee, T.C., Hsiung, P.A.: Mutation coverage estimation for model checking. In: Automated Technology for Verification and Analysis (ATVA), Taipei, Taiwan (2004)
6. McMillan, K.L.: Applying SAT methods in unbounded symbolic model checking. In: Proceedings of the International Conference on Computer Aided Verification (CAV), Copenhagen, Denmark (2002)
7. Nguyen, M.D., Thalmaier, M., Wedler, M., Bormann, J., Stoffel, D., Kunz, W.: Unbounded protocol compliance verification using interval property checking with invariants. IEEE Trans. Comput. Aided Des. 27(11), 2068–2082 (2008)
8. Onespin Solutions GmbH: Germany. OneSpin 360MV. http://www.onespin-solutions.com
9. Schickel, M., Nimbler, V., Braun, M., Eveking, H.: On consistency and completeness of property sets: exploiting the property-based design process. In: Proceedings of Forum on Design Languages, Darmstadt, Germany (2006)
10. Spear, C.: SystemVerilog for Verification: A Guide to Learning the Testbench Language Features. Springer, Dordrecht, New York (2008)
11. Urdahl, J., Stoffel, D., Bormann, J., Wedler, M., Kunz, W.: Path predicate abstraction by complete interval property checking. In: Proceedings of the International Conference on Formal Methods in Computer-Aided Design (FMCAD), Lugano, Switzerland, pp. 207–215 (2010)
12. Yuan, J., Pixley, C., Aziz, A.: Constraint-Based Verification. Springer, New York (2006)

Chapter 2
Efficient Refinement Strategy Exploiting Component Properties in a CEGAR Process

Syed Hussein S. Alwi, Cécile Braunstein, and Emmanuelle Encrenaz

Abstract Embedded systems are usually composed of several components and in practice, these components generally have been independently verified to ensure that they respect their specifications before being integrated into a larger system. Therefore, we would like to exploit the specification (i.e. verified CTL properties) of the components in the objective of verifying a global property of the system. A complete concrete system may not be directly verifiable due to the state explosion problem, thus abstraction and eventually refinement process are required. In this paper, we propose a technique to select properties in order to generate a good abstraction and reduce refinement iterations. We have conducted several preliminary experimentations which show that our approach is promising in comparison to other abstraction-refinement techniques implemented in VIS [1].

2.1 Introduction

The embedded systems correspond to the integration into the same electronic circuit, a huge number of complex functionalities performed by several heterogeneous components. Current SoC (System on Chips) contain multiple processors executing numerous cooperating tasks, specialized co-processors (for particular data treatment or communication purposes), Radio-Frequency components, etc. These systems are usually submitted to safety and robustness requirements. Depending on their application domains, their failure may induce serious damages and catastrophic consequences.

Therefore, it is important to ensure, during their design phase, their correctness with respect to their specifications. Errors found late in the design of these systems

---

S.H.S. Alwi (✉) • C. Braunstein • E. Encrenaz
Université Pierre et Marie Curie Paris 6, LIP6-SOC (CNRS UMR 7606),
4, place Jussieu, 75005 Paris, France
e-mail: syed-hussein.alwi@lip6.fr; cecile.braunstein@lip6.fr; emmanuelle.encrenaz@lip6.fr

J. Haase (ed.), *Models, Methods, and Tools for Complex Chip Design*, Lecture Notes in Electrical Engineering 265, DOI 10.1007/978-3-319-01418-0_2,
© Springer International Publishing Switzerland 2014

is a major problem for electronic circuit designers and programmers as it may delay getting a new product to the market or cause failure of some critical devices that are already in use. System verification using formal methods such as model checking guarantees a high level of quality in terms of safety and reliability while reducing financial risk.

The main challenge in model checking is dealing with the state space combinatorial explosion phenomenon. A strategy to overcome the state explosion problem is by performing abstraction. A method for the construction of an abstract state graph of an arbitrary system automatically was first proposed by Graf and Saidi [2] using Pvs theorem prover. Here, the abstract states are generated from the valuations of a set of predicates on the concrete variables. The construction approach is automatic and incremental.

In 2000, an interesting abstraction-refinement methodology called counterexample guided abstraction refinement (CEGAR) was proposed by Clarke and al. [3]. The abstraction was done by generating an abstract model of the system by considering only the variables that possibly have a role in verifying a particular property. In this technique, the counterexample provided by the model-checker in case of failure is used to refine the system.

Several tools using counterexample-guided abstraction refinement technique, like those implemented in the VIS model-checker, have been developed such as SLAM, a software model-checker by Microsoft Research [4], BLAST (Berkeley Lazy Abstraction Software Verification Tool), a software model-checker for C programs [5] and VCEGAR (Verilog Counterexample Guided Abstraction Refinement), a hardware model-checker which performs verification at the RTL (Register Transfer Language) level [6]. However, relying on counterexamples generated by the model checker as the only source for refinement may not be conclusive.

Recently, a CEGAR based technique that combines precise and approximated methods within one abstraction-refinement loop was proposed for software verification [7]. This technique uses predicate abstraction and provides a strategy that interleaves approximated abstraction which is fast to compute and precise abstraction which is slow. The result shows a good compromise between the number of refinement iterations and verification time.

An alternative method to get over the state explosion problem is the compositional strategy. The strategy is based on the assume-guarantee reasoning where assumptions are made on other components of the systems when verifying one component. Several works have manipulated this technique notably in [8] where Grumberg and Long described the methodology using a subset of CTL in their framework and later in [9] where Henzinger and al. presented their successful implementations and case study regarding this approach.

Xie and Browne have proposed a method for software verification based on composition of several components [10]. Their main objective is developing components that could be reused with certitude that their behaviors will always respect their specification when associated in a proper composition. Therefore, temporal properties of the software are specified, verified and packaged with the component for possible reuse. The implementation of this approach on software has been successful and the application of the assume-guarantee reasoning has considerably reduced the model checking complexity. A comprehensive approach to model-check component-based systems with abstraction refinement technique that uses verified properties as abstractions has been presented in [11].

In [12], Peng, Mokhtari and Tahar have presented a possible implementation of assume-guarantee approach where the specifications are in ACTL. Moreover, they managed to perform the synthetisation of the ACTL formulas into Verilog HDL behavior level program. The synthesized program can be used to check properties that the system's components must guarantee. Since, there have been other works on construction of components from interval temporal logic properties which could be used to speed up verification process [13, 14].

In 2007, a method to build abstractions of components into AKS (Abstract Kripke Structure), based on the set of the properties (CTL) each component verifies was presented in [15]. The method is actually a tentative to associate compositional and abstraction-refinement verification techniques. The generations of AKS from CTL formula have been successfully automated [16]. This work will be the base of the techniques in this paper.

Contribution: In this paper we present a strategy to exploit the properties of verified component in the goal of verifying complex systems with a good initial abstraction and eventually being conclusive in a small number of refinement iterations. We propose a technique to classify component properties according to their pertinency towards the global property, thus, enabling a better selection of properties for the initial abstraction generation. Furthermore, in the case where the verification is not conclusive, we propose a technique guided by the counterexample given by the model-checker to select supplementary properties to improve the abstraction.

In the next section, we will give an overview of our framework and introduce the notations that will be used later. The rest of the paper is organized as follows: Sect. 2.3 details our strategy of refinement. Section 2.4 presents the experimentation results and finally, Sect. 2.5 draws the conclusions and summarize our possible future works.

2.2 Our Framework

The model-checking technique we propose is based on the Counterexample-guided Abstraction Refinement (CEGAR) methodology [3]. The overall description of our methodology is shown in Fig. 2.1. We take into account the structure of the system as a set of synchronous components, each of which has been previously verified and a set of CTL properties is attached to each component. This set

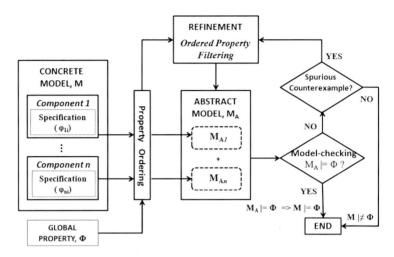

Fig. 2.1 Verification process

refers to the specification of the component. We would like to verify whether a concrete model, M presumably big sized and composed of several components, satisfies a global ACTL property Φ. Instead of building the product of the concrete components, we replace each concrete component by an abstraction of its behavior derived from a subset of the CTL properties it satisfies. Each abstract component represents an over-approximation of the set of behaviors of its related concrete component [15].

As shown in [17] for over-approximation abstraction, if Φ holds in the abstract model then it holds in the concrete model as well. However, if Φ does not hold in the abstract model then one cannot conclude anything regarding the concrete model until the counterexample has been analyzed. The test of spurious counter-example is then translated into a SAT problem as in [3]. When a counterexample is proven to be spurious, the refinement phase occurs, injecting more preciseness into the (abstract) model to be analyzed.

2.2.1 Concrete System Definition

As mentioned earlier, our concrete model consists of several components and each component comes with its specification. The concrete system is a synchronous composition of components, each of which described as a Moore machine.

Definition 2.1. A *Moore machine* C is defined by a tuple $\langle I, O, R, \delta, \lambda, \mathbf{R}_0 \rangle$, where,

- I is a finite set of Boolean input signals.
- O is a finite set of Boolean output signals.

2 Efficient Refinement Strategy Exploiting Component Properties

- R is a finite set of Boolean sequential elements (registers).
- $\delta : 2^I \times 2^R \to 2^R$ is the transition function.
- $\lambda : 2^R \to 2^O$ is the output function.
- $\mathbf{R}_0 \subseteq 2^R$ is the set of initial states.

States (or configurations) of the circuit correspond to Boolean configurations of all the sequential elements.

Definition 2.2. A *Concrete system* M is obtained by synchronous composition of the component.
$M = C_1 \parallel C_2 \parallel \ldots \parallel C_n$, where each C_i is a Moore machine with a specification associated $\varphi_i = \{\varphi_i^1 \ldots \varphi_i^k\}$. Each φ_i^j is a CTL\X formula whose propositions AP belong to $\{I_i \cup O_i \cup R_i\}$.

2.2.2 Abstraction Definition

Our abstraction reduces the size of the representation model by letting free some of its variables. The point is to determine the good set of variable to be freed and when to free them. We take advantage of the CTL specification of each component: a CTL property may be seen as a partial view of the tree of behaviors of its variables configuration. All the variables not specified by the property can be freed. We introduced the Abstract Kripke Structure (AKS for short) which exactly specifies when the variable of the property can be freed. The abstraction of a component is represented by an AKS, derived from a subset of the CTL properties the component satisfies. Roughly speaking, AKS(φ), the AKS derived from a CTL property φ, simulates all execution trees whose initial state satisfies φ. In AKS(φ), states are tagged with the truth values of φ's atomic propositions, among the four truth values of Belnap's logic [18]: inconsistent ($\perp$), false (**f**), true (**t**) and unknown ($\top$). States with inconsistent truth values are not represented since they refer to non possible assignments of the atomic propositions. A set of fairness constraints eliminates non-progress cycles. The transformation algorithm of a CTL\X property into an AKS is described in [15, 19].

Definition 2.3. Given a CTL\X property φ whose set of atomic propositions is AP, an *Abstract Kripke Structure*, $AKS(\varphi) = (AP, \hat{S}, \hat{S}_0, \hat{L}, \hat{R}, \hat{F})$ is a 6-tuple consisting of:

- AP: The finite set of atomic propositions of property φ
- $\hat{S}$: a finite set of states
- $\hat{S}_0 \subseteq \hat{S}$: a set of initial states
- $\hat{L} : \hat{S} \to \mathscr{B}^{|AP|}$ with $\mathscr{B} = \{\perp, \mathbf{f}, \mathbf{t}, \top\}$: a labeling function which labels each state with configuration of current value of each atomic proposition.
- $\hat{R} \subseteq \hat{S} \times \hat{S}$: a transition relation where $\forall s \in \hat{S}, \exists s' \in \hat{S}$ such that $(s, s') \in \hat{R}$
- $\hat{F}$: a set of fairness constraints (generalized Büchi acceptance condition)

We denote by $\hat{L}(s)$, the configuration of atomic propositions in state s, and by $\hat{L}(s)[p]$, the projection of configuration $\hat{L}(s)$ according to atomic proposition p.

As the abstract model $\hat{M}$ is generated from the conjunction of verified properties of the components in the concrete model M, it can be seen as the composition of the AKS of each property. The AKS composition has been defined in [19]; it extends the classical synchronous composition of Moore machine to deal with four-valued variables.

Definition 2.4. An *Abstract model* $\hat{M}$ is obtained by synchronous composition of components abstractions. Let n be the number of components in the model and m be the number of selected verified properties of a component; let C_j be a component of the concrete model M and φ_j^k is a CTL formula describing a satisfied property of component C_j. Let $AKS(\varphi_{C_j^k})$ the AKS generated from φ_j^k. We have $\forall j \in [1,n]$ and $\forall k \in [1,m]$:

- $\hat{C}_j = AKS(\varphi_{C_j^1}) \,\|\, AKS(\varphi_{C_j^2}) \,\|\ldots\|\, AKS(\varphi_{C_j^k}) \,\|\ldots\| \, AKS(\varphi_{C_j^m})$
- $\hat{M} = \hat{C}_1 \,\|\, \hat{C}_2 \,\|\, \ldots \,\|\, \hat{C}_j \,\|\, \ldots \,\|\, \hat{C}_n$

In an AKS, a state where a variable p is *unknown* can simulate all states in which p is either true or false. It is a concise representation of the set of more concrete states in which p is either true or false. A state s is said to be an *abstract state* if one of its variable p is *unknown*.

Definition 2.5. The *concretization* of an abstract state s with respect to the variable p (*unknown* in that state), assigns either true or false to p.
The *abstraction* of a state s with respect to the variable p (either true or false in that state), assigns *unknown* to p.

Property 2.1 (Concretization). Let A_i and A_j two abstractions such that A_j is obtained by concretizing one abstract variable of A_i (resp. A_i is obtained by abstracting one variable in A_j). Then A_i simulates A_j, denoted by $A_j \sqsubseteq A_i$.

Proof. As the concretization of state reduces the set of concrete configuration the abstract state represents but does not affect the transition relation of the AKS. The unroll execution tree of A_j is a sub-tree of the one of A_i. Then A_i simulates A_j. ☐

Property 2.2 (Composition and Concretization). Let $\hat{M}_i$ be an abstract model of M and φ_j^k be a property of a component C_j of M, $\hat{M}_{i+1} = \hat{M}_i \,\|\, AKS(\varphi_j^k)$ is more concrete that $\hat{M}_i$, $\hat{M}_{i+1} \sqsubseteq \hat{M}_i$.

Proof. Let $s = (s_i, s_{\varphi_j^k})$ be a state in S_{i+1}, such that $s_i \in S_i$ and $s_{\varphi_j^k} \in S_{\varphi_j^k}$. The label of s_{i+1} is obtained by applying the Belnap's logic operators *and* to the four-valued values of variables in s_i and $s_{\varphi_j^k}$. For all $p \in AP_i \cup AP_{\varphi_j^k}$ we have the following label :

- $\hat{L}_{i+1}[p] = \top$ iff p is *unknown* in both states or does not belong to the set of atomic proposition.

2 Efficient Refinement Strategy Exploiting Component Properties 23

- $\hat{L}_{i+1}[p] = \mathbf{t}$ (or $\mathbf{f}$) iff p is true (or false) in $s_{\varphi_j^k}$ (resp. s_i) and *unknown* in s_i (resp. $s_{\varphi_j^k}$).

By Property 2.1, $\hat{M}_{i+1}$ is more concrete than $\hat{M}_i$ and by the property of parallel composition, $\hat{M}_i \sqsubseteq \hat{M}_i \parallel AKS(\varphi_j^k)$. $\square$

2.2.3 Initial Abstraction

Given a global property Φ, the property to be verified by the composition of the concrete components model, an abstract model is generated by selecting some of the properties of the components which are relevant to Φ. In the initial abstraction generation, all variables that appear in Φ have to be represented. Therefore the properties in the specification of each component where these variables are present will be used to generate the initial abstraction, $\hat{M}_0$ and we will verify the satisfiability of the global property Φ on this abstract model. If the model-checking failed and the counterexample given is found to be spurious, we will then proceed with the refinement process.

2.3 Refinement

2.3.1 Properties of Good Refinement

When a counterexample is found to be spurious, it means that the current abstract model $\hat{M}_i$ is too coarse and has to be refined. In this section, we will discuss about the refinement technique based on the integration of more verified properties of the concrete model's components in the abstract model to be generated. Moreover, the refinement step from $\hat{M}_i$ to $\hat{M}_{i+1}$ respects the properties below:

Definition 2.6. An efficient *refinement* verifies the following properties:

1. The new refinement is an over-approximation of the concrete model: $\hat{M} \sqsubseteq \hat{M}_{i+1}$.
2. The new refinement is more concrete than the previous one: $\hat{M}_{i+1} \sqsubseteq \hat{M}_i$.
3. The spurious counterexample in $\hat{M}_i$ is removed from $\hat{M}_{i+1}$.

Furthermore, the refinement steps should be easy to compute and ensure a fast convergence by minimizing the number of iterations of the CEGAR loop.

Refinements based on the concretization of selected abstract variables in $\hat{M}_i$ ensure Item 2. Concretization can be performed by modifying the AKS of $\hat{M}_i$ by changing some abstract value to concrete ones. However, this approach is rude: in order to ensure Item 1, the concretization needs to be consistent with the sequences of values in the concrete system. The difficulty resides in defining the proper abstract variable to concretize, at which precise instant, and with which Boolean value.

We propose to compose the abstraction with another AKS to build a good refinement according to Definition 2.6. We have several options. The most straightforward method consists in building an AKS representing all possible executions except the spurious counterexample; however the AKS representation may be huge and the process is not guaranteed to converge. A second possibility is to build an AKS with additional CTL properties of the components; the AKS remains small but Item 3 is not guaranteed, hence delaying the convergence. The final proposal combines both previous ones: first local CTL properties eliminating the spurious counterexample are determined, and then the corresponding AKS is synchronized with the one of $\hat{M}_i$.

2.3.2 Negation of the Counterexample

The counterexample at a refinement step i, σ, is a path in the abstract model $\hat{M}_i$ which dissatisfies Φ. In the counterexample given by the model-checker, the variable configuration in each state is Boolean. We name $\hat{L}_i$ this new labeling. The spurious counterexample σ is defined such that:

Definition 2.7. Let σ be a *spurious counterexample* in $\hat{M}_i = \langle AP_i, \hat{S}_i, \hat{S}_{0i}, \hat{L}_i, \hat{R}_i, \hat{F}_i \rangle$ of length $|\sigma| = n$: $\sigma = s_0 \to s_1 \ldots \to s_n$ with $(s_k, s_{k+1}) \in \hat{R}_i \ \forall k \in [0..n-1]$.

- All its variables are concrete: $\forall s_i$ and $\forall p \in AP_i$, p is either true or false according to $\hat{L}_i$. (not *unknown*), and s_0 is an initial state of the concrete system: $s_0 \in \mathbf{R}_0$
- σ is a counterexample in $\hat{M}_i$: $s_0 \not\models \Phi$.
- σ is not a path of the concrete system M: $\exists k \in [1..n-1]$ such that $\forall j < k, (s_j, s_{j+1}) \in R$ and $(s_k, s_{k+1}) \notin R$.

The construction of the AKS representing all executions except the one described by the spurious counterexample is done in two steps.

2.3.2.1 Step 1: Build the Structure of the AKS

Definition 2.8. Let σ be a spurious counterexampleof length $|\sigma| = n$, the *AKS of the counterexample negation $AKS(\overline{\sigma}) = \langle AP_{\overline{\sigma}}, \hat{S}_{\overline{\sigma}}, \hat{S}_{0\overline{\sigma}}, \hat{L}_{\overline{\sigma}}, \hat{R}_{\overline{\sigma}}, \hat{F}_{\overline{\sigma}} \rangle$* is such that:

- $AP_{\overline{\sigma}} = AP_i$: The set of atomic propositions coincides with the one of σ
- $\hat{S}_{\overline{\sigma}}$: $\{s_T\} \cup \{s_i' | \forall i \in [0..n-2] \land s_i \in \sigma\} \cup \{\bar{s}_i | \forall i \in [0..n-1] \land s_i \in \sigma\}$
- $\hat{L}_{\overline{\sigma}}$ with $L_{\overline{\sigma}}(s_i') = L_i(s_i), \forall i \in [0..n-2]$ and $L(s_T) = \{\top, \forall p \in AP_{\overline{\sigma}}\}$, $L_{\overline{\sigma}}(\bar{s}_i)$ is explained in the next construction step.
- $\hat{S}_{0\overline{\sigma}} = \{s_0', \bar{s}_0\}$
- $\hat{R}_{\overline{\sigma}} = \{(\bar{s}_i, s_T), \forall i \in [0..n-1]\} \cup \{(s_i', \bar{s}_{i+1}), \forall i \in [0..n-2]\} \cup \{(s_i', s_{i+1}'), \forall i \in [0..n-3]\}$
- $\hat{F}_{\overline{\sigma}} = \emptyset$

The labeling function of s_i' represents (concrete) configuration of state s_i and state $\bar{s}_i$ represents all configurations *but* the one of s_i. This last set may not be

2 Efficient Refinement Strategy Exploiting Component Properties

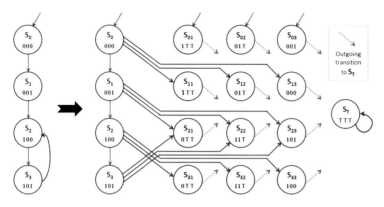

Fig. 2.2 An example of a negation of the counterexample AKS, $AKS(\sigma)$

representable by the labeling function defined in Definition 2.3. State labeling is treated in the second step. s_T is a state where all atomic propositions are *unknown*.

2.3.2.2 Step 2: Expand State Configurations Representing the Negation of a Concrete Configuration

The set of configurations associated with a state $\bar{s}_i$ represents the negation of the one represented by $L_i(s_i)$. This negation is not representable by the label of a single state but rather by a union of $|AP|$ labels.

Example: Assume $AP = \{v_0, v_1, v_2\}$ and $\sigma = s_0 \rightarrow s_1$ and $\hat{L}(s_0) = \{\mathbf{f}, \mathbf{f}, \mathbf{f}\}$ the configuration associated with s_0 assigns false to each variable. The negation of this configuration represents a set of seven concrete configurations which are covered by three (abstract) configurations: $\{\{\mathbf{t}, \top, \top\}, \{\mathbf{f}, \mathbf{t}, \top\}, \{\mathbf{f}, \mathbf{f}, \mathbf{t}\}\}$.

To build the final AKS representing all sequences but spurious counterexample σ, one replaces in $AKS(\bar{\sigma})$ each state $\bar{s}_i$ by $k = |AP_{\bar{\sigma}}|$ states $\bar{s}_i^j$ with $j \in [0..k-1]$ and assigns to each of them a label of k variables $\{v_0, \ldots, v_{k-1}\}$ defined such that: $\hat{L}(\bar{s}_i^j) = \{\forall l \in [0..j-1], v_l = L_i(s_i)[v_l]; v_j = \neg L_i(s_i)[v_j]; \forall l \in [j+1..k-1], v_l = \top\}$. Each state $\bar{s}_i^j$ is connected to the same predecessor and successor states as state $\bar{s}_i$. This final AKS presents a number of states in $\mathcal{O}(|\sigma| \times |AP|)$.

Figure 2.2 shows an example of the negation of a counterexample AKS built from a counterexample $\sigma = s_0 \rightarrow s_1 \rightarrow s_2 \rightarrow s_3 \rightarrow s_2$. The counterexample consists of four states with a loop to a previous state. The negation of the counterexample AKS allows all possible behaviors except the last step in σ. Therefore, the complementary states of every state in the counterexample are presented and at any step, a state in σ can proceed to these complementary states. The elimination of the last step is obtained by forcing its predecessor to the complementary states of the last step. All complementary states then leads to the terminal state, S_T which represents all possible behaviors in the future steps.

2.3.2.3 Reduction of the Negation of the Counterexample AKS

In the AKS generated, the set of configurations associated to the negation of a counterexample state may be redundant i.e. some configurations are represented several times in the AKS. Furthermore, all the states in negation part of the AKS have a unique successor namely the S_T state. Therefore, in the objective to reduce the number of states, these counterexample negation states with identical variable configurations can be merged. The merge definitions to generate the negation of the counterexample reduced AKS are given below.

Definition 2.9. *Merge condition:* Let $AKS(\bar{\sigma}) = (AP_{\bar{\sigma}}, \hat{S}_{\bar{\sigma}}, \hat{S}_{0\bar{\sigma}}, \hat{L}_{\bar{\sigma}}, \hat{R}_{\bar{\sigma}}, \hat{F}_{\bar{\sigma}})$. s_1 and s_2 are two counterexample negation states in M: $(s_1, s_2) \in \hat{S}_{\bar{\sigma}} \setminus \{s_T, s \in \sigma\}$. s_1 and s_2 can be merged iff

$$\hat{L}_{\bar{\sigma}}(s_1) = \hat{L}_{\bar{\sigma}}(s_2)$$

Definition 2.10. *Merging action:* Let $AKS(\bar{\sigma}) = (AP_{\bar{\sigma}}, \hat{S}_{\bar{\sigma}}, \hat{S}_{0\bar{\sigma}}, \hat{L}_{\bar{\sigma}}, \hat{R}_{\bar{\sigma}}, \hat{F}_{\bar{\sigma}})$ and its reduced AKS, $AKS(\bar{\sigma})' = (AP'_{\bar{\sigma}}, \hat{S}'_{\bar{\sigma}}, \hat{S}'_{0\bar{\sigma}}, \hat{L}'_{\bar{\sigma}}, \hat{R}'_{\bar{\sigma}}, \hat{F}'_{\bar{\sigma}})$ applying the Definition 2.9.

$s' \in \hat{S}', \forall (s_1, s_2) \in \hat{S} \setminus \{s_T, s \in \sigma\}, s' = \text{merge } (s_1, s_2) \Rightarrow$

- $\hat{L}'_{\bar{\sigma}}(s') = \hat{L}_{\bar{\sigma}}(s_1) = \hat{L}_{\bar{\sigma}}(s_2)$
- $\forall ((s_{p1}, s_1), (s_{p2}, s_2)) \in \hat{R}^2, ((s_{p1}, s'), (s_{p2}, s')) \in \hat{R}'^2$
- $\forall ((s_1, s_{s1}), (s_2, s_{s2})) \in \hat{R}^2, ((s', s_{s1}), (s', s_{s2})) \in \hat{R}'^2$

Property 2.3. $AKS(\bar{\sigma})'$ and $AKS(\bar{\sigma})$ are bisimulation-equivalent:

$$AKS(\bar{\sigma})' \sim AKS(\bar{\sigma})$$

Proof. Let $AKS(\bar{\sigma}) = (AP_{\bar{\sigma}}, \hat{S}_{\bar{\sigma}}, \hat{S}_{0\bar{\sigma}}, \hat{L}_{\bar{\sigma}}, \hat{R}_{\bar{\sigma}}, \hat{F}_{\bar{\sigma}})$ and its reduced AKS,
$AKS(\bar{\sigma})' = (AP'_{\bar{\sigma}}, \hat{S}'_{\bar{\sigma}}, \hat{S}'_{0\bar{\sigma}}, \hat{L}'_{\bar{\sigma}}, \hat{R}'_{\bar{\sigma}}, \hat{F}'_{\bar{\sigma}})$.
All the initial states in $\hat{S}_{0\bar{\sigma}}$ are represented in $\hat{S}'_{0\bar{\sigma}}$ and vice versa. $\forall (s_1, s_2) \in \hat{R}_{\bar{\sigma}}, \exists (s'_1, s'_2) \in \hat{R}'_{\bar{\sigma}}$ where $\hat{L}_{\bar{\sigma}}(s_i) = \hat{L}'_{\bar{\sigma}}(s'_i)$, and the other way around is also true. Therefore, $AKS(\bar{\sigma})' \sim AKS(\bar{\sigma})$. $\qquad\square$

Figure 2.3 demonstrates the gain from the reduction process on the generation of the negation of the counterexample AKS from the counterexample σ in the previous example. In the Fig. 2.3 above, we can see that all the complementary states have a unique variable configuration and the duplicates no longer present in the AKS. This simplification technique helps to reduce the size of the system without having a degradation in terms of property verification as the resulted AKS is bisimilar to the original one. Even though the gain may seem insignificant at first sight, the reduction done may be precious when the technique in conducted on many refinement iterations. Therefore, this reduction technique will be applied systematically on this method of refinement.

However, removing, at each refinement step, the spurious counterexample *only* induces a low convergence. Moreover, in some cases, this strategy may not converge: suppose that all sequences of the form $a.b^*.c$ are spurious counterexamples (here a, b and c represent concrete state configurations). Assume, at a given

2 Efficient Refinement Strategy Exploiting Component Properties

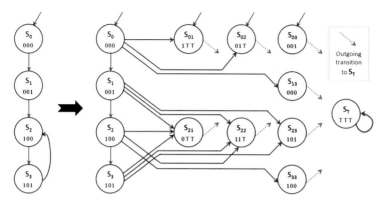

Fig. 2.3 An example of a reduced negation of the counterexample AKS, $AKS(\sigma)'$

refinement step i, a particular counterexample $\sigma_i = s_0 \to s_1 \to \ldots s_n$ with $L(s_0) = a, \forall k \in [1, n-1], L(s_k) = b, L(s_n) = c$. Removing this counterexample does not prevent from a new spurious counterexample at step $i+1$: $\sigma_{i+1} = s_0 \to s_1 \to \ldots s_{n+1}$ with $L(s_0) = a, \forall k \in [1, n], L(s_k) = b, L(s_{n+1}) = c$. The strategy consisting of elimination spurious counterexample *one by one* diverges in this case. Furthermore, we cannot eliminate all the sequences of the form $a.b^*.c$ in a unique refinement step since we do not a priori know if at least one of these sequences is executable in the concrete model.

Therefore, from these considerations, we are interested in removing *sets of behaviors encompassing the spurious counterexample* while still guaranteeing an over-approximation of the set of tree-organized behaviors of the concrete model. The strengthening of the abstraction $\hat{M}_i$ with the addition of AKS of already verified local CTL properties eliminates sets of behaviors and guarantees the over-approximation (Property 2.2) but does not guarantee the elimination of the counterexample. We present in the following section a strategy to select sets of CTL properties eliminating the spurious counterexample.

2.3.3 Ordering of Properties

We propose a heuristic to order the properties depending on the structure of each component. In order to do so, the variable dependency of the variables present in global property has to be analyzed. After this point, we refer to the variables present in the global property as *primary variables*.

We observed that the closer a variable is to the primary variable, the higher influence it has on it. Moreover, a global property often specifies the behavior at the interface of components. Typically, a global property ensures that a message sent is always acknowledged or the good target gets the message. This kind of behavior relates the input-output behaviors of components. We have decided to allocate an

28 S.H.S. Alwi et al.

extra weight for interface variables whereas variables which do not interfere with a primary variable are weighted 0. Here is how we proceed:

1. Build the structural dependency graph for all primary variables.
2. Compute the depth of all variables in all dependency graphs. Note that a variable may belong to more than one dependency graph, in that case we consider the minimum depth.
3. Give a weight to each variable (see Algorithm 1).
4. Compute the weight of properties for each component: sum of the property variables weight.

Algorithm 1: Compute weight

Input: G, the set of all dependency graph variable
V, the set of variables
Output: $\{(v,w)|v \in V, w \in \mathbb{N}\}$, The set of variables with their weight
1 **begin**
2    $p = \max(\text{depth}(G))$
3   **for** $v \in V$ **do**
4      $d = \text{depth}(v)$;
5      $w = 2^{p-d} * p$;
6   **if** $d == 0$ **then** v is primary variable
7        $w = 5 * w$;
8   **end**
9   **if** $v \in I \cup O$ **then** v is an interface variable
10        $w = 3 * w$
11   **end**
12   **end**
13 **end**

The Algorithm 1 gives weight according to the variable distance to the primary variable with extra weight for interface variable and primary variable. It is definitely not an exact pertinence calculation of properties but provides a good indicator of their possible impact on the global property. After this pre-processing phase, we have a list of properties ordered according to their pertinence with regards to the global property.

2.3.3.1 Example

In this example, we have a global property $\phi = A((p = 1)U(q = 1))$; which consists of two primary variables: p and q. As shown in Fig. 2.4, the primary variable p is dependent of three the other variables: x, y and z whereas the primary variable q is dependent of four variables: r, s, u and v. The maximum depth of between the two primary variables dependency graphs is three ($q \leftarrow r \leftarrow u \leftarrow v$). Furthermore, apart from p and q being the primary variables, we have y, z, s and v which are interface

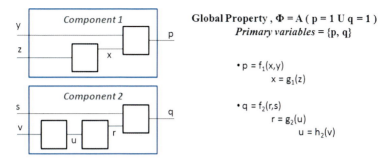

Fig. 2.4 Example of variable dependency

variables. Let's assume that we have a set of properties that includes $\varphi_a - \varphi_g$, with the weight computation algorithm given previously, the property φ_a which consists of variables p, y and z will therefore obtain the highest total weight and the rest of the properties will be ordered as follows:

List of ordered component properties:

1. $\varphi_a(p, y, z)$
2. $\varphi_b(q, s, v)$
3. $\varphi_c(p, y)$
4. $\varphi_d(q, r, v)$
5. $\varphi_e(p, z)$
6. $\varphi_f(x, z)$
7. $\varphi_g(r, u, v)$
8. ...

Here we can see that the top property φ_a only consists of primary variable p, therefore the highest property in the list containing q will also be selected in the initial abstraction generation.

Selected properties for the initial abstraction:

1. $\varphi_a(p, y, z)$
2. $\varphi_b(q, s, v)$

2.3.4 Filtering Properties

The refinement step consists of adding new AKS of properties selected according to their pertinence. As we would like to ensure the elimination of the counterexample previously found, we filter out properties that do not have an impact on the counterexample σ thus will not eliminate it. In order to reach this objective, a Abstract Kripke structure of the counterexample σ, $K(\sigma)$ is generated. $K(\sigma)$ is a succession of states corresponding to the counterexample path which dissatisfies the global property Φ.

Definition 2.11. Let σ be a counterexample of length n in $\hat{M}_i$ such that $\sigma = s_0 \to s_1 \to \ldots \to s_{n-1}$. The *Kripke structure derived from* σ is 6-tuple $K(\sigma_i) = (AP_\sigma, S_\sigma, S_{0\sigma}, L_\sigma, R_\sigma, F_\sigma)$ such that:

- $AP_\sigma = AP_i$: a finite set of atomic propositions which corresponds to the variables in the abstract model
- $S_\sigma = \{s_i | s_i \in \sigma\} \cup \{s_T\}$
- $S_{0\sigma} = \{s_0\}$
- $L_\sigma = \hat{L}_i \cup L(s_T) = \{\top, \forall p \in AP_\sigma\}$
- $R_\sigma = \{(s_k, s_{k+1}) | (s_k \to s_{k+1}) \in \sigma\} \cup \{(s_{n-1}, s_T)\}$
- $F_\sigma = \emptyset$

All the properties available for refinement are then model-checked on $K(\sigma)$. If the property holds then the property will not eliminate the counterexample. Hence this property is not a good candidate for refinement. Therefore the highest weighted property not satisfied in $K(\sigma)$ is chosen to be integrated in the next refinement step. This process is iterated for each refinement step.

Property 2.4. **Counterexample eviction**

1. If $K(\sigma) \models \varphi \Rightarrow AKS(\varphi)$ will not eliminate σ.
2. If $K(\sigma) \nvDash \varphi \Rightarrow AKS(\varphi)$ will eliminate σ.

Proof. 1. By construction, $AKS(\varphi)$ simulates all models that verify φ. Thus the tree described by $K(\sigma)$ is simulated by $AKS(\varphi)$, it implies that σ is still a possible path in $AKS(\varphi)$.
2. $K(\sigma)$, where φ does not hold, is not simulated by $AKS(\varphi)$, thus σ is not a possible path in $AKS(\varphi)$ otherwise $AKS(\varphi) \nvDash \varphi$ that is not feasible due to AKS definition and the composition with M_i with $AKS(\varphi)$ will eliminate σ. □

The proposed approach ensures that the refinement excludes the counterexample and respects the Definition 2.6. We will show in our experiments that first, the time needed to build an AKS is negligible and secondly the refinement converges rapidly.

2.4 Experimental Results

We have conducted preliminary experiments to test and compare the performance of our strategy with existing techniques available in VIS. There are several abstraction-refinement techniques implemented in VIS accessible via *approximate_model_check*, *iterative_model_check*, *check_invariant* and *incremental_ctl_verification* commands. However, among the available techniques, *incremental_ctl_verification* is the only one that supports CTL formulas and fairness constraints which are necessary in our test platforms. It is an automatic abstraction

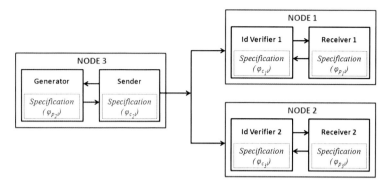

Fig. 2.5 CAN protocol platform

Table 2.1 Statistics on the VCI-PI and CAN bus platform

Experiment platform			Number of BDD variables	BDD size	Number of reachable states	Analysis time (s)
VCI-PI	Concrete model	1 master-1 slave	304	7,207	4.711e+3	6.36
		2 masters-1 slave	445	24,406	7.71723e+06	35.2
		4 masters-1 slave	721	84,118	3.17332e+12	2,818.3
		4 masters-2 slaves	895	238,990	5.708e+15	68,882.3[a]
	Final abstract model for ϕ_1	1 master-1 slave	197	76	5.03316e+07	0.01
		2 masters-1 slave	301	99	4.12317e+11	0.02
		4 masters-1 slave	501	147	3.45876e+18	0.03
		4 masters-2 slaves	589	167	7.08355e+21	0.04
	Final abstract model for ϕ_2	1 master-1 slave	194	50	2.62144e+07	0
		2 masters-1 slave	298	73	2.14748e+11	0.01
		4 masters-1 slave	498	121	1.80144e+18	0.02
		4 masters-2 slaves	586	141	3.68935e+21	0.02
CAN bus	Concrete model		822	161,730	3.7354e+07	300.12
	Final abstract model for ϕ_3		425	187	1.66005e+12	0.03
	Final abstract model for ϕ_4		425	187	1.66005e+12	0.04

[a] Computed on a calculation server: 2x Xeon X5650, 72 Go RAM

refinement algorithm which generates an initial conservative abstraction principally by reducing the size of the latches by a constant factor. If the initial abstraction is not conclusive, a *goal set* will then be computed in order to guide the refinement process [20, 21].

We have executed and compared the execution time and the number of refinement iterations for two examples: VCI-PI platform consisting of Virtual Component Interface (VCI), a PI-Bus and VCI-PI protocol converter and a simplified version of a CAN bus platform consisting of three nodes on a CAN bus as shown in Fig. 2.5. Table 2.1 gives the size and the statistics concerning the VCI-PI platform and CAN bus platform verified. All the values are obtained using the *compute_reach*

Table 2.2 Verification Results

Experiment platform	Global property	Verification technique	Refinement iteration	Verification time (s)
		Property selection	1	2.2
VCI-PI:	ϕ_1	Incremental	0	6.3
1 master		Standard MC	–	6.06
–		Property selection	0	1.0
1 slave	ϕ_2	Incremental	562	200.9
		Standard MC	–	6.13
		Property selection	1	2.0
VCI-PI:	ϕ_1	Incremental	0	20.4
2 masters		Standard MC	–	37.9
–		Property selection	0	1.0
1 slave	ϕ_2	Incremental	74	786.3
		Standard MC	–	39.4
		Property selection	1	2.1
VCI-PI:	ϕ_1	Incremental	0	261.6
4 masters		Standard MC	–	>1 day
–		Property selection	0	1.0
1 slave	ϕ_2	Incremental	0	263.5
		Standard MC	–	>1 day
		Property selection	1	2.2
VCI-PI:	ϕ_1	Incremental	N/A	>1 day
4 masters		Standard MC	–	>1 day
–		Property selection	0	1.1
2 slaves	ϕ_2	Incremental	N/A	>1 day
		Standard MC	–	>1 day
		Property selection	0	1.02
	ϕ_3	Incremental	N/A	>1 day
CAN		Standard MC	–	2,645.4
bus		Property selection	0	1.01
	ϕ_4	Incremental	N/A	>1 day
		Standard MC	–	1,678.1

command with option -v 1 in VIS except the number of BDD variables, computed using the *print_bdd_stats* command. The experiments have been executed on a PC with an AMD Athlon dual-core processor 4450e and 1.8 GB of RAM memory.

In Table 2.2, we compare the execution time and the number of refinement between our technique (Prop. Select.), *incremental_ctl_verification* (Incremental) and the standard model checking (Standard MC) computed using the *model_check* command in VIS (Note: Dynamic variable ordering has been enabled with sift method). For the VCI-PI platform, the global property ϕ_1 is the type $AF((p = 1) * AF(q = 1))$ and ϕ_2 is actually a stronger version of the same formula with $AG(AF((p = 1) * AF(q = 1)))$ where all requests to write on the PI-Bus will finally be granted in the future. We have a total of 26 verified components properties to be selected in the VCI-PI platform. In comparison to ϕ_2, we can see that, a better set of properties available will result in a better abstraction and less refinement iterations.

In the case of the CAN bus platform, the global property ϕ_3 is the type $AG(((p' = 1) * (q' = 1) * AF(r_1 = 1)) \rightarrow AF((s_1 = 1) * AF(t_1 = 1)))$ and $\phi_4 = AG(((p' = 1) * (q' = 1) * AG(r_2 = 0)) \rightarrow AG((s_2 = 0) * (t_2 = 0)))$. They describe the correct transmission of generated messages to the receivers. We have at our disposal 103 verified component properties and after the selection process for the initial abstraction, 3 selected component properties were sufficient to verify both global properties without refinement.

Globally, we can see that our technique, for these examples, systematically computes faster than the other two methods and interestingly in the case where the size of the platform increases by adding more connected components, in contrary to the other two methods, our computation time remains stable. This is mainly due to the fact that for small number of properties, our abstraction is generated almost instantly and as only pertinent properties are selected, not many refinement iterations are required in order to complete the verification process. It is also important to note that the properties tested are simple and we have in our property selection list the local properties required to satisfy the global property.

2.5 Negation of the Counterexample as a Complementary Strategy

A well constituted specification is a prerequisite for an efficient refinement strategy based on property selection technique. However, in practice, we don't always have at our disposal a complete specification. Hence, it may be possible that at a particular refinement iteration, none of the properties available is capable of eliminating the counterexample. In this case, we propose the negation of the counterexample technique as a complementary strategy.

Let's suppose that there are no properties available to refine our CAN Bus abstract model for the verification of a global property $\phi_5 = A((a = 1)U((a = 0) * AX((b_1 = 1) * (b_2 = 1))))$; where b_1 and b_2 are outputs of the Receiver 1 and 2 respectively and they are set to 1 in the next step after the signal $a = 0$ is on the bus which indicates the start of frame. As our current AKS generator is only capable of generating $CTL \setminus X$ properties only, the initial abstractions of each component were built with the aid of the AF operators which allows more satisfaction configurations than the AX operator.

Therefore, in this example, the negation of the counterexample strategy could help to eliminate the different configurations that are present in the abstraction. The first counterexample σ_1 provided by the model checker gives the undesired configuration where the output b_1 still remains at 0 right after $a = 0$. Thus, the negation of the counterexample is applied on this counterexample configuration to eliminate it.

Table 2.3 Statistics at each refinement step with the negation of counterexample technique

Iteration	Number of BDD variables	BDD size	Number of reachable states	Model checking result
0	424	199	4.00015e + 09	$\widehat{M_0} \nvDash \phi_5$
1	426	199	3.79804e + 09	$\widehat{M_1} \nvDash \phi_5$
2	428	249	3.6633e + 09	$\widehat{M_2} \vDash \phi_5$

In the following iteration, the model-checker provides a rather similar configuration of undesired behavior with this time the output b_2 which remains at 0 after $a = 0$. As previously done, the negation of counterexample is applied on this counterexample σ_2. Finally, after these two refinement iterations, the abstract model built managed to satisfy the property ϕ_5. Table 2.3 shows some statistics at each refinement iteration.

2.6 Conclusion and Future Works

We have presented a new strategy in the abstraction generation and refinement which is well adapted for compositional embedded systems. This verification technique is compatible and suits well in the natural development process of complex systems. Our preliminary experimental results show an interesting performance in terms of duration of abstraction generation and the number of refinement iteration. Moreover, this technique enables us to overcome repetitive counterexamples due to the presence of cycles in the system's graph.

Nevertheless, in order to function well, this refinement technique requires a well constituted specification of every components of the concrete model. Furthermore, it may be possible that none of the properties available is capable of eliminating the counterexample which is probably due to an incomplete specification or a counterexample that should be eliminated by the product of local properties.

We have also demonstrated a possible application of the negation of the counterexample technique as a complementary strategy albeit limited to certain form of counterexamples only. Indeed, the negation of counterexample technique is inefficient when dealing with counterexample with a cycle in the prefix (e.g. $a.b^*.c$).

In this case, other refinement techniques such as the identification of a good set of local properties to be integrated simultaneously should be considered. We are currently investigating other complementary techniques to overcome these particular cases. The work of Kroening [22], for example, could also help us in improving the specification of the model: at the component level, or for groups of components.

Furthermore, we are also examining a comprehensive new strategy that exploits the finite-state machines (FSMs) of the components in the verification process. A procedure to generate properties which are directly derived from the component's FSM structure is considered as a solution to overcome the insufficiency of component properties to be selected for the abstraction generation. These ongoing researches will enrich the existing verification techniques in property based abstraction generation.

Acknowledgements We thank Neha Agarwal for the implementation of the negation of the counterexample (without reduction) AKS generator which is the base of the generator with reduction used in this paper.

References

1. The VIS Group: VIS: A system for verification and synthesis, In: Alur, R., Henzinger, T.A. (eds.) Proceedings of the 8th International Conference, CAV '96, New Brunswick. LNCS, vol. 1102, pp. 428–432. Springer, Berlin/Heidelberg (1996)
2. Graf, S., Saïdi, H.: Construction of abstract state graphs with PVS. In: Grumberg, O. (ed.) Computer Aided Verification (CAV '97), Haifa. LNCS, vol. 1254. Springer, London, Springer Berlin Heidelberg (1997)
3. Clarke, E.M., Grumberg, O., Jha, S., Lu, Y., Veith, H.: Counterexample-guided abstraction refinement. In: CAV'00, Chicago. LNCS (2000)
4. Ball, T., Cook, B., Levin, V., Rajamani, S.K.: SLAM and static driver verifier: Technology transfer of formal methods inside microsoft. In: 4th International Conference on Integrated Formal Methods, Canterbury, vol. 2999, pp. 1–20. Springer (2004)
5. Beyer, D., Henzinger, T.A., Jhala, R., Majumdar, R.: The software model checker blast: Applications to software engineering. Int. J. Softw. Tools Technol. Trans. 9(5–6), 505–525 (2007)
6. Jain, H., Kroening, D., Sharygina, N., Clarke, E.: VCEGAR: Verilog counterexample guided abstraction refinement. In: TACAS'07, Braga, 2007
7. Sharygina, N., Tonetta, S., Tsitovich, A.: An abstraction refinement approach combining precise and approximated techniques. Int. J. Softw. Tools Technol. Trans. 14, 1–14 (2012)
8. Grumberg, O., Long, D.E.: Model checking and modular verification. In: International Conference on Concurrency Theory, Amsterdam, vol. 527, pp. 250–263. Springer, Berlin/Heidelberg (1991)
9. Henzinger, T.A., Qadeer, S., Rajamani, S.K.: You assume, we guarantee: Methodology and case studies. In: CAV '98, Vancouver, vol. 1427, pp. 440–451. Springer, Berlin/Heidelberg (1998)
10. Xie, F., Browne, J.C.: Verified systems by composition from verified components, in ES-EC/FSE 2003. In: 11th ACM SIGSOFT Symposium on Foundations of Software Engineering Conference, Helsinki, pp. 227–286. ACM (2003)
11. Li, J., Sun, X., Xie, F., Song, X.: Component-based abstraction refinement. In: 10th International Conference on Software Reuse (ICSR), Beijing, pp. 39–51. Springer (2008)
12. Peng, H., Mokhtari, Y., Tahar, S.: Environment synthesis for compositional model checking. In: ICCD '02: 20th International Conference on Computer Design, Freiburg, pp. 70–75. IEEE Computer Society (2002)
13. Schickel, M., Nimbler, V., Braun, M., Eveking, H.: On consistency and completeness of property-sets: Exploiting the property-based design process. In: FDL '06: Forum on Specification and Design Languages, Darmstadt (2006)
14. Nguyen, M.D., Wedler, M., Stoffel, D., Kunz, W.: Formal hardware/software co-verification by interval property checking with abstraction. In: Design Automation Conference (DAC'11), San Diego 2011

15. Braunstein, C., Encrenaz, E.: Using CTL formulae as component abstraction in a design verification flow. In: ACSD, Bratislava, pp. 80–89. IEEE Computer Society (2007)
16. Bara, A.: Abstraction de Composant pour la Vérification par Model-Checking, Mémoire de Diplôme Universitaire OMP – LIP6-SOC, (2008)
17. Clarke, E.M., Grumberg, O., Long, D.E.: Model checking and abstraction. ACM Trans. Program. Lang. Syst. **16**(5), 1512–1542 (1994)
18. Belnap, N.: A useful four-valued logic. In: Modern Uses of Multiple-Valued Logic, pp. 8–37. Springer, Berlin/Heidelberg (1977)
19. Braunstein, C.: Conception Incrémentale, Vérification de Composants Matériels et Méthode d'abstraction pour la Vérification de Systèmes Intégrés sur Puce. Ph.D. thesis, Université Pierre et Marie Curie, LIP6-SOC (2007)
20. Pardo, S., Hachtel, G.: Incremental CTL model checking using BDD subsetting. In: DAC '98: 35th Design Automation Conference, San Francisco, pp. 457–462. ACM (1998)
21. Pardo, S., Hachtel, G.: Automatic abstraction technique for propositional mu-Calculus model checking. In: CAV '97, Haifa, vol. 1254, pp. 12–23. Springer, (1997)
22. Purandare, M., Wahl, T., Kroening, D.: Strengthening properties using abstraction refinement. In: Proceedings of DATE '09, Nice, pp. 1692–1697. ACM (2009)

Chapter 3
Formal Specification Level

Rolf Drechsler, Mathias Soeken, and Robert Wille

Abstract The steadily increasing complexity of the design of embedded systems led to the development of both an elaborated design flow that includes various abstraction levels and corresponding methods for synthesis and verification. However, until today the initial system specification is provided in natural language which is manually translated into a formal implementation e.g. at the *Electronic System Level* (ESL) by means of SystemC in a time-consuming and error-prone process.

In this chapter, we envision a design flow which incorporates a *Formal Specification Level* (FSL) aiming at bridging the gap between the informal textbook specification and the formal ESL implementation. Modeling languages such as UML or SysML are envisaged for this purpose. Recent accomplishments towards this envisioned design flow, namely the automatic derivation of formal models from natural language descriptions, verification of formal models in the absence of an implementation, and code generation techniques, are briefly reviewed.

3.1 Introduction

Being composed of up to several billion components, the design of embedded systems is one of the most complex problems people are facing today. While it was possible to fully design such systems gate by gate on the drawing table 40 years ago, this procedure has become intractable due to the ever increasing complexity. As a consequence, elaborated design flows have been developed over the last decades in which several levels of abstraction are considered.

R. Drechsler (✉) • M. Soeken • R. Wille
Group of Computer Architecture, University of Bremen, Bremen, Germany

Cyber-Physical Systems, DFKI GmbH, Bremen, Germany
e-mail: drechsle@informatik.unibremen.de; msoeken@informatik.unibremen.de; rwille@informatik.unibremen.de

J. Haase (ed.), *Models, Methods, and Tools for Complex Chip Design*, Lecture Notes in Electrical Engineering 265, DOI 10.1007/978-3-319-01418-0_3,
© Springer International Publishing Switzerland 2014

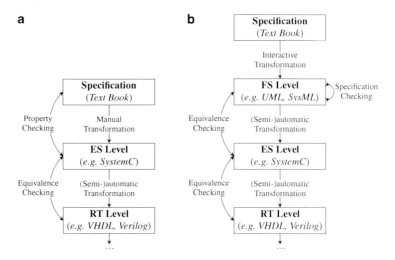

Fig. 3.1 Conventional and envisioned design flow. (**a**) Conventional design flow. (**b**) Envisioned design flow

Today, a design flow as briefly illustrated in Fig. 3.1 is applied. The initial starting point is given by means of a specification which is usually provided in terms of a text book description, however, in order to perform even the simplest automatic synthesis techniques, a formal representation of the specification is required. For this purpose, an initial implementation is generated at the *Electronic System Level* (ESL) using high-level programming languages such as SystemC. This system level description enables the execution and simulation of the desired design, but still hides details concerning a precise realization in both hardware and software. From this description, the system model is consecutively refined in successive steps leading to descriptions at the *Register Transfer Level* (RTL), the *gate level*, and the *physical level*. At the end of this process, the resulting chip is sent to a chip manufacturer.

As embedded systems are often employed in safety critical systems such as avionic, automotive, and medical applications, ensuring the correctness is of high importance. For this purpose, usually each transformation from one abstraction level to the next refinement is checked for equivalence. But due to the absence of a formal description at the specification level, automatic verification techniques are not applicable for the comparison with the system level. Moreover, as the system level representation is manually derived from the textual specification, this step is particularly prone to errors and mistakes.

So far property checking is applied to address this issue by extracting properties from the specification in terms of temporal and modal logic expressions which can subsequently be checked with algorithms known as model checkers [6]. Further techniques called coverage detection exist, which can automatically determine if enough properties have been written, i.e. whether the full behavior is considered by all properties [10, 12]. However, the main obstacle remains: the specification is provided in natural language and a formal representation needs to be manually derived from it for further processing. Motivated by this, researchers started working on closing the gap between the informal textbook specifications and the respective ESL implementation [9, 13, 26].

In this work, we envision a new design flow that exploits recent achievements in this area and we propose two major extensions. First, we follow the steady strive for higher levels of abstraction and enrich the specification itself by formal descriptions. Modeling languages such as the *Unified Modeling Language* (UML, [23]) or the *System Modeling Language* (SysML, [29]) combined with constraints provided in the *Object Constraint Language* (OCL, [28]) provide proper syntax and semantic for this purpose.[1] While models in those languages means remain abstract enough for the specification level, their formal description enables (semi-)automatic verification and code construction. As a result, crucial design flaws can already be detected at the specification level in the absence of a precise implementation.

Second, initial solutions (e.g. [26]) are utilized which allow to automatically derive UML/OCL descriptions from natural language specifications. Achievements in the area of natural language processing [15] and knowledge representation [21] are exploited for this purpose. In fact, already simple grammatical analyses enable e.g. the derivation of

1. Basic components of a system (which can be derived from nouns in a sentence),
2. Their functions (which can be derived from verbs in a sentence), and
3. Attributes (which can be derived from adjectives in a sentence).

Having such methods, we envision a design flow which includes a *Formal Specification Level* (FSL) as shown in Fig. 3.1. This flow enables to (semi-)automatically derive formal models from a given specification provided in natural language. Formal methods are applied to this description to verify the correctness of the design prior to an implementation. If all checks passed, code skeletons for synthesis and formal properties for verification are extracted for further usage within the remaining stages of the established design flow.

In the remainder of this chapter, the general ideas and first accomplishments towards this envisioned design flow are presented. The following section briefly introduces the necessary background to keep the chapter self-contained. Afterwards, Sect. 3.3 outlines the proposed extension to the overall design flow in detail. The respective steps for mapping a natural language specification to a formal model, checking the correctness of that formal model, and transforming the formal model into an implementation are then outlined in Sects. 3.4–3.6, respectively. Finally, remaining challenges to be addressed are discussed and the chapter is concluded in Sect. 3.8.

---

[1]In the following, we focus on UML in combination with OCL, while the general concepts are similarly applicable to other modeling languages as well.

3.2 Preliminaries

In this work, the *Unified Modeling Language* (UML) is applied to represent the code skeletons and test cases which are semi-automatically derived from natural language. Besides that, we also exploit language processing tools. To keep the paper self-contained, the underlying concepts of UML and the applied tools are briefly reviewed in the following.

3.2.1 Unified Modeling Language

A UML *class diagram* is used to represent the structure of a system. The main component of a class diagram is a *class* that describes an atomic entity of the model. A class itself consists of *attributes* and *operations*. Attributes describe the information which is stored in the class (e.g. member variables). Operations define possible actions that can be executed e.g. in order to modify the values of attributes. Classes can be set into relation via *associations*. The type of a relation is expressed by *multiplicities* that are added to each *association end*. Class diagrams can be extended by constraints in the *Object Constraint Language* (OCL) such as invariants that further restrict the attribute values.

Example 3.1. Figure 3.2 illustrates the specification of a simple computer architecture in UML using the class diagram notation. The structure of the system is defined by means of four classes, namely a *Processor*, a *Kernel*, a *Thread*, and a *Memory*. Attributes such as *maxCapacity* provide further details on the respective components (e.g. the maximal capacity of the processor). An operation *spawn* is defined on the class *Processor*. An invariant states that the number of a processor's thread must not exceed the CPU's capacity.

A detailed overview of the UML is provided in [23].

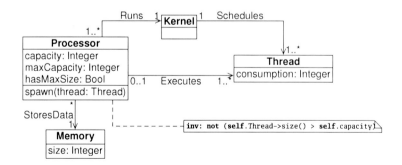

Fig. 3.2 UML class diagram

3 Formal Specification Level   41

1. `<noun.person>` **waiter, server** — *(a person whose occupation is to serve at table (as in a restaurant))*
2. `<noun.person>` **server1** — *((court games) the player who serves to start a point)*
3. `<noun.artifact>` **server1, host** — *((computer science) a computer that provides client stations with access to files and printers as shared resources to a computer network)*
4. `<noun.artifact>` **server** — *(utensil used in serving food or drink)*

Fig. 3.3 WordNet output of the query for "server"

3.2.2 Natural Language Processing

3.2.2.1 Word Sense Disambiguation

When the correct sense of a word must be determined, *Word Sense Disambiguation* (WSD, [17]) is applied. Given e.g. the sentence "The server delivers the website," it is easy for a human to identify "server" as a technical device. For a computer, in contrast, it is an impossible task to determine the correct sense with no additional information. Thinking in terms of an MDE context, it is not clear whether "server" describes a class which is part of the model or an actor that interacts with the model.

In case of ambiguity, WordNet [21] is used for dictionary-based word sense disambiguation [15] in the scope of this work. WordNet is a lexical dictionary of English, consisting of more than 90,000 word senses and 166,000 pairs connecting senses with a semantic meaning. WordNet is designed to be used by external programs and for many senses it also provides example sentences. Figure 3.3 displays the results of a WordNet query for the word "server".

In this work, WordNet is applied to determine the semantics of the sentences in a specification.

3.2.2.2 Constituency Grammars

A constituency grammar [5] is used to decompose a sentence into its constituent parts, usually depicted as a *phrase structure tree* (PST, cf. Fig. 3.4). A PST is a tree whose root is labeled with the most general phrase structure, in case of Fig. 3.4 it is labeled *S* referring to the whole sentence. The leaves of the tree are the words of the sentence. Moving along the branches from the root to the leaves, the vertices become more specialized phrase structures. Following the leftmost branch from the example shown in Fig. 3.4 we get the following structures: *S* (sentence) → *NP* (noun phrase) → *DT* (determiner) → "The" (word within the sentence). The parent of a leaf corresponds to the part-of-speech (POS, or tag) of the leaf, i.e. "The" and "the" are determiners, "server" and "website" are nouns and "delivers" is a verb. For details on how a PST is extracted from a sentence we refer to [15] and [4].

In this work, the constituency grammars are applied to process the structure of the sentences in a specification.

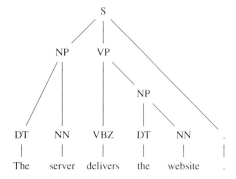

Fig. 3.4 Phrase structure tree (PST) of the sentence "The server delivers the website."

3.2.2.3 Dependency Grammars

In order to represent dependencies between individual words, natural language processing techniques make use of *dependency parses* [7], i.e. binary dependency relations are extracted from the sentences. As an example the relation *nsubj* binds a verb to its subject. The usual notation for this relation is *relation(governor,dependent)*, e.g. *nsubj(delivers,server)*. To avoid ambiguities, the position of the word within the sentence can be appended to the words of the relation (i.e. *nsubj (delivers-3, server-2)*, cf. Fig. 3.5).

The typed dependencies of a sentence s can be understood as an edge-labeled graph whose vertices represent words and labels the type of the dependency. There is an edge $g \xrightarrow{r} d$ if and only if $r(g,d)$ is a dependency of s. For a visualization cf. Fig. 3.5.

For example, the nouns are assigned their articles using the *det* relation. The relations *nsubj* and *dobj* allow for the extraction of the typical *subject-verb-object* form. In Fig. 3.5, the verb *delivers* is connected to the subject *server* and to the object *website*.

In this work, dependency parses are applied to extract formal properties from informal requirements.

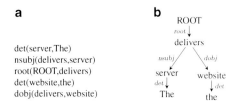

Fig. 3.5 Typed dependencies of the sentence "The server delivers the website." (a) List of typed dependencies. (b) Typed dependencies as a directed graph. Edge labels denote the type of the relation between the words

3.3 Formal Specification Level

Figure 3.6 provides a more detailed view on the proposed extension for the envisioned design flow. The main goal is to (semi-)automatically derive an ESL-implementation in SystemC[2] from a (textbook) specification provided in natural language. Given natural language test cases and requirements from the specification, an initial SystemC implementation, an executable testbench for simulation, and operation contracts (pre- and post-conditions as motivated by *Design-by-Contract* [20]) are (semi-)automatically generated. For this purpose, the Formal Specification Level as shown in Fig. 3.1 and detailed in Fig. 3.6 is introduced as a new abstraction level which includes three stages.

In the first stage (cf. Sect. 3.4), the test cases and the requirements are mapped from their natural language description into a formal representation by means of UML/OCL. NLP techniques are exploited in order to extract the desired information. More precisely, the following steps are conducted in this first stage:

- *Determine the structure of the design*
  Using e.g. a grammatical analysis, the basic components of the considered system are derived from the natural language specification. From the resulting information, a UML class diagram is created which provides a first formal description of the structure for the considered design.
- *Determine the properties of the design*
  After the structure has been obtained, the requirements of the specification can be considered in detail. From them, formal properties which need to be satisfied by the design are derived and represented in terms of OCL expressions.

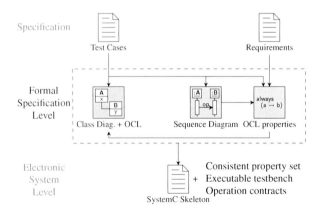

Fig. 3.6 Overview of the formal specification level

---

[2]Note that SystemC is just a representative for any high-level object-oriented hardware description language and can readily be replaced.

As a result, the first stage leads to a formal description of the desired system in terms of UML/OCL.

In the second stage (cf. Sect. 3.5), this formal description is used to conduct initial checks for correctness. This includes consistency checks such as checking whether it is possible to instantiate the desired system considering all constraints and requirements, but also first behavioral checks such as checking whether it is possible to reach a prohibited state. This allows for the detection of design flaws already in very early design steps in the absence of a precise implementation.

In the third stage (cf. Sect. 3.6), after all checks have passed and no errors have been determined, a skeleton for the system level implementation as well as corresponding testbenches are derived.

In the next sections, first accomplishments with respect to these stages of the FSL are illustrated.

3.4 Mapping Natural Language Specifications to the Formal Specification Level

The first stage addresses the (semi-)automatic determination of a formal representation describing the structure and the properties of a system that is specified in natural language. First accomplishments for the former two aspects have been presented in [26] and are reviewed in the following two sub-sections. Afterwards, ideas on the property determination are presented.

3.4.1 Determine the Structure of the Design

Technical specifications are often written in a very specific way: often a strong focus is put on using simple short sentences which contain information. From these sentences, much information can already been determined automatically. As an example, consider the following excerpt of a specification based on the example class diagram given in Fig. 3.2.

> A processor spawns a simple thread.
> The number of a processor's threads must not exceed the processor's capacity.

Figure 3.7 illustrates that already from these two sentences a significant amount of structural information can be extracted: Since processor and thread are common nouns, it can be concluded that they represent components of the considered system (to be represented by classes). Preceded adjectives (such as simple) substantiate objects and thus shall be added as attributes to the corresponding class. Verbs correlate to operations which can be invoked by components or actors. Moreover, prepositions help to determine relations between classes. For example, a processor's threads implies a relation.

3 Formal Specification Level

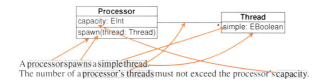

Fig. 3.7 Determine the structure of the design

Recent progress in the development of NLP technologies enables to extract much of these information in a (semi-)automatic manner. More precisely, NLP parsers (e.g. the one presented in [18]) are able to decompose a sentence in terms of a phrase structure tree which assigns each atomic word to a syntactic word type (such as noun, verb, or adjective) and also groups words into larger sub-parts of the sentence (cf. Sect. 3.2).

However, sometimes the syntactical and grammatical information alone is not sufficient. For example in the first sentence from Fig. 3.7, two nouns are identified in the PST, i.e. processor and thread. To determine whether a noun describes a class or an actor interacting with the system, additional information needs to be derived. For this purpose, we can make use of word sense disambiguation. Dictionaries such as WordNet categorize words into lexicographic files. Based on this categories, a classifier can be implemented: If a noun is assigned to the category *person*, the noun likely refers to an actor, if a noun is assigned to the category *artifact* or *object*, the noun rather refers to a part of the system.

Although making use of word sense disambiguation, a clear assignment is not always possible. As an example, the noun processor can both refer to a person or to an artifact. Moreover, WordNet assigns the different synsets with frequency counts which give an indication of how frequent a word is commonly used. This allows to guess the correct sense, but in the case of processor both senses have the same frequency count. User interaction is required to resolve this ambiguity.

Overall, exploiting these NLP technologies, a UML class diagram formally representing the structure of the considered system can automatically be determined in many cases. However, since the textual description always can contain ambiguities, manual interactions with the design engineer cannot entirely be excluded leading to a (semi-)automatic and assisted approach as evaluated in [26].

3.4.2 Determine the Properties of the Design

When a model is available, e.g. determined using methods described in the previous section, formal behavioral specifications can be extracted from English sentences using the approach presented in this section. More precisely, the goal is to support the designer in creating a formal specification in OCL from given informal natural language requirements. During the generation process, it is exploited, that despite

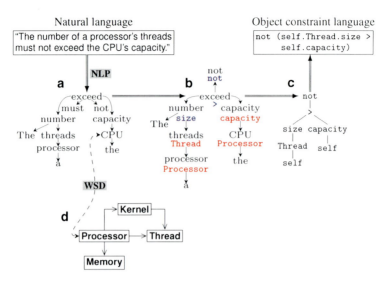

Fig. 3.8 Extracting OCL constraints

the undoubtedly existing differences, the given input (i.e. the sentence in natural language) and the desired output (i.e. the formal requirement in OCL) are indeed quite similar. While this is often not evident in a direct comparison, structural analyses unveil the correlation between the input and the output. This is illustrated in the following example.

Consider the informal requirement "The number of a processor's threads must not exceed the CPU's capacity" and its formal counterpart

```
not (self.Thread->size() > self.capacity).
```

A direct mapping of these two descriptions (cf. the boxes at the top of Fig. 3.8) is not straightforward. However, after a prior application of semantical and grammatical analyses followed by a normalization, a promising representation can be obtained as shown in Fig. 3.8a–c. In fact, the resulting normalized dependency graph of the sentence (cf. Fig. 3.8b) is almost identical to the resulting abstract syntax tree (AST) of the OCL constraint (cf. Fig. 3.8c).

However, the example in Fig. 3.8 also shows that, due to the wide scope of natural language, a direct mapping of *all* parts of the informal requirement to the appropriate identifier or OCL operations is not guaranteed. Often different grammatical forms of words (e.g. due to declension or conjugation) or the use of synonymous descriptions represent obstacles to a one-to-one mapping from the dependency graph to the AST. Dictionary-based word sense disambiguation can be applied to address these problems. Using this technique, normal forms and synonymous identifiers are determined.

With respect to the example in Fig. 3.8, while e.g. *not* and *capacity* can easily be mapped from the dependency graph to the corresponding OCL expression (highlighted in blue color) or model element (highlighted in red color), respectively, a correct mapping of *CPU* is not obvious at a first glance. However, the application of WSD unveils that among others the word "CPU" is a synonym for "processor" (for a visualization of the WSD process cf. Fig. 3.8d). Since *Processor* is a class in the model, it can be assumed that "CPU" is just an alternative description of "processor" in the informal description. Hence, substituting both words does not affect the meaning of the requirement, but enables a correct mapping from the informal requirement to the formal requirement.

Further approaches which help to derive formal properties from natural language requirements have been proposed in [19] and [8]. In [19], a systematic approach is suggested in which sub-sentences of the requirement are manually mapped into formal properties. This allows for a guided and thus less error-prone manual translation. Furthermore, reoccurring sub-sentences can be re-used.

Besides that, it is also possible to generalize properties from test cases when they obey a certain structure. For example, in the context of *Behavior Driven Development* (BDD) the structure of a test case is often given by a *Given A, When B, Then C* template [22]. Each of these sentences is linked to test code which should be executed when running the test cases. Since *A* corresponds to environment constraints, *B* corresponds to the antecedent, and *C* corresponds to the consequent of a property, formal properties can be generalized from such test cases [8].

3.5 Checking Correctness at the Formal Specification Level

Having passed the first stage of extracting information from natural language descriptions, a formal specification composed of the structure and properties of the system is available. In the second stage, this representation is being used as a basis for initial correctness checks targeting conceptual questions such as whether the system is free of any contradictions which would disallow an ongoing implementation. For this purpose, approaches presented in [2, 11, 14, 24] for static verification, presented in [27] for invariant elimination, and presented in [3, 25] for dynamic verification can be applied. In [30], also first debugging approaches have been introduced.

3.5.1 Verification of Static Aspects

Having a formal representation of the design does not necessarily imply that a working implementation can be generated from it. In fact, the formal model may inherit constraints which contradict each other. As a result, no valid instantiation would be possible and any implementation would be erroneous from scratch. The FSL enables to detect such errors before any code is written.

Approaches introduced e.g. in [2, 11, 14, 24] can be utilized for this purpose. They take the obtained UML diagram (representing the structure) together with the properties (which are encoded as OCL invariants) and automatically perform the above described *consistency checks*. Besides enumerative methods [11], also elaborated formal approaches have been proposed in the recent past [24]. Considering the abstract description of the models (usually, no complex data-structures are applied), particularly the latter approaches are applicable to quite significantly complex designs.

3.5.2 Invariant Removal

At the FSL, invariants are a proper description mean to represent properties the design has to satisfy. However, when it comes to verification they may cause unnecessary overhead. Since invariants are assumed globally, i.e. for each possible system state of the system, they have to be considered all the time. Even if only a certain functionality of a design is under verification, *invariants* of the entire model have to be assumed additionally.

An alternative to prevent this overhead has been proposed in [27]. Here, invariants are iteratively removed and replaced instead with a smaller set of pre- and post-conditions for certain operations. This enables to entirely eliminate all invariants without changing the semantics of the model. Since additionally, pre- and post-conditions only have to be considered locally when the corresponding function is called, this reduces the overhead.

Furthermore, invariant elimination enables a design flow in which the implementation of different operations can be conducted by different developers. Then, the respective sub-teams do not have to globally consider all the invariants anymore, but just the local pre- and post-conditions of the corresponding operation.

3.5.3 Verification of Dynamic Aspects

Finally, also the dynamic behavior can be verified at the FSL. This is possible due to the above-mentioned pre- and post-conditions of operations which enable a descriptive representation of the behavior, without giving a precise implementation. A pre-condition describes in which states an operation can be called, while the post-condition describes the effect an operation has on that system state. These conditions may be specified directly from the designer or are determined by the invariant elimination step described above.

Any model where its operations are enriched with pre- and post-conditions can be transformed into an instance similar to *Bounded Model Checking* (BMC) [1] and, therefore, allows for addressing certain dynamic verification tasks. In fact, similar to verification at the implementation level, operation sequences can be determined that lead e.g. to bad states, good states, live locks, or dead locks [25]. Utilizing these techniques, again, errors can be detected before any code is written.

3.6 Mapping from Formal Specification Level to the Electronic System Level

Finally, the formally modeled and verified design shall be implemented in an ESL language so that it can be further refined using the established conventional design flow. Also in this final stage of the FSL the formal representation can be exploited.

In fact, the corresponding UML/OCL descriptions allow for a generation of code parts for the implementation process. This includes

- Code stubs generated from class diagrams (Sect. 3.4.1),
- Generalized properties from the parameterized test cases (Sect. 3.4.2),
- A consistent property set (Sect. 3.5.1),
- And contracts for the operations of a class (Sect. 3.5.2).

Further, the verification of dynamic aspects plays a significant role in the transition from the FSL to the ESL. As briefly discussed in Sect. 3.5.3, all dynamic aspects, i.e. the interaction of the components, can be checked in the absence of a precise implementation. As an example, it can be ensured that the model is deadlock-free or that all operations can be reached from given initial states. Hence, after the implementation phase, it is sufficient to check whether the implementation of each single operation adheres correctly to its contracts. That is, assuming the pre-condition and executing the code must imply the post-conditions. Since the verification of the operations can be performed locally without considering the whole system, verification effort can be decreased.

3.7 Tool Support

The first stage of the proposed design flow, i.e. the mapping from natural language specifications into formal models, has been implemented into the IDE lips [16]. The IDE considers natural language as a *first class citizen* in the design flow of systems and software. Just like modeling languages such as SysML are used for describing the model and programming languages such as Java are used for writing the implementation, natural language is used to write the specification. Many IDE concepts used for easing the use with modeling and programming language can be exploited in order to ease the writing of the specification as well. Figure 3.9 shows a

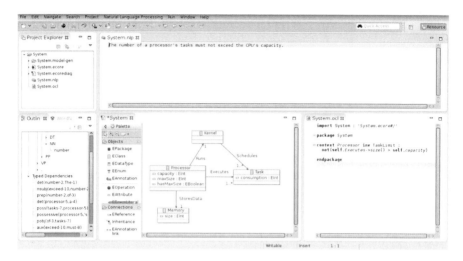

Fig. 3.9 lips IDE

screenshot with the example sentence, the underlying model, and the resulting OCL constraint. Furthermore, it can be seen that e.g. the outline is used in order to show the phrase structure and its dependencies.

3.8 Conclusion

In this paper, we envisioned a new design flow which includes an FSL representing the desired design using modeling languages such as UML or SysML combined with constraints provided in languages such as OCL. The proposed flow bridges the gap between the natural language textbook specification and the formal ESL implementation. We illustrated that first accomplishments towards the envisioned design flow have already been made: NLP techniques are available to derive formal descriptions of natural language specifications, verification approaches based on modeling languages allow to detect design errors prior to a precise implementation, and code generation techniques can be applied to generate code stubs, executable testbenches, etc.

Acknowledgements This work was supported by the German Research Foundation (DFG) within the Reinhart Koselleck project DR 287/23-1. The authors would like to thank Melanie Diepenbeck, Daniel Große, Ulrich Kühne, Hoang M. Le, and Julia Seiter for interesting and helpful discussions.

References

1. Biere, A., Cimatti, A., Clarke, E.M., Strichman, O., Zhu, Y.: Bounded model checking. Adv. Comput. **58**, 117–148 (2003)
2. Cabot, J., Clarisó, R., Riera, D.: Verification of UML/OCL class diagrams using constraint programming. In: IEEE International Conference on Software Testing Verification and Validation Workshop, Lillehammer, pp. 73–80 (2008)
3. Cabot, J., Clarisó, R., Riera, D.: Verifying UML/OCL operation contracts. In: Leuschel, M., Wehrheim, H. (eds.) Integrated Formal Methods. Lecture Notes in Computer Science, vol. 5423, pp. 40–55. Springer, Berlin/New York (2009)
4. Carnie, A.: Syntax: A Generative Introduction. Blackwell, Malden (2007)
5. Chomsky, N.: Three models for the description of language. IRE Trans. Inf. Theory **2**(3), 113–124 (1956)
6. Clarke, E.M., Jr., Grumberg, O., Peled, D.A.: Model Checking. MIT, Cambridge (1999)
7. de Marneffe, M.C., MacCartney, B., Manning, C.D.: Generating typed dependency parses from phrase structure parses. In: Conference on Language Resources and Evaluation, Genoa, pp. 449–454 (2006)
8. Diepenbeck, M., Soeken, M., Große, D., Drechsler, R.: Behavior driven development for circuit design and verification. In: IEEE International High Level Design Validation and Test Workshop, Huntington Beach (2012)
9. Drechsler, R.: Quality-driven design of embedded systems based on specification in natural language. In: EUROMICRO Symposium on Digital System Design, Oulu (2011)
10. Drechsler, R., Diepenbeck, M., Große, D., Kühne, U., Le, H.M., Seiter, J., Soeken, M., Wille, R.: Completeness-driven development. In: ICGT, Bremen, pp. 38–50 (2012)
11. Gogolla, M., Kuhlmann, M., Hamann, L.: Consistency, independence and consequences in UML and OCL models. In: Tests and Proofs, pp. 90–104. Springer, Berlin/New York (2009)
12. Große, D., Drechsler, R.: Quality-Driven SystemC Design. Springer, Dordrecht/Heidelberg/London/New York (2009)
13. Harris, I.G.: Extracting design information from natural language specifications. In: Design Automation Conference, San Francisco, pp. 1256–1257 (2012)
14. Jackson, D.: Software Abstractions: Logic, Language, and Analysis. MIT, Cambridge (2006)
15. Jurafsky, D., Martin, J.H.: Speech and Language Processing. Pearson Prentice Hall, Upper Saddle River (2008)
16. Keszocze, O., Soeken, M., Kuksa, E., Drechsler, R.: lips: An IDE for model driven engineering based on natural language processing. In: Workshop on Natural Language Analysis in Software Engineering, San Francisco (2013)
17. Kilgarriff, A., Rosenzweig, J.: Framework and results for english SENSEVAL. Comput. Humanit. **34**, 15–48 (2000)
18. Klein, D., Manning, C.D.: Accurate unlexicalized parsing. In: Annual Meeting of the Association for Computational Linguistics, Sapporo, pp. 423–430 (2003)
19. Le, H.M., Große, D., Drechsler, R.: From requirements and scenarios to ESL design in systemC. In: International Symposium on Electronic System Design, Kolkata (2012)
20. Meyer, B., Nerson, J.M., Matsuo, M.: EIFFEL: object-oriented design for software engineering. In: Nichols, H.K., Simpson, D. (eds.) European Software Engineering Conference, Strasbourg. Lecture Notes in Computer Science, vol. 289, pp. 221–229. Springer (1987)
21. Miller, G.A.: WordNet: a lexical database for English. Commun. ACM **38**(11), 39–41 (1995)
22. North, D.: Behavior modification: the evolution of behavior-driven development. Better Softw. **8**(3) (2006)
23. Rumbaugh, J., Jacobson, I., Booch, G.: The Unified Modeling Language Reference Manual. Addison-Wesley Longman, Essex (1999)
24. Soeken, M., Wille, R., Kuhlmann, M., Gogolla, M., Drechsler, R.: Verifying UML/OCL models using Boolean satisfiability. In: Design, Automation and Test in Europe, Dresden, pp. 1341–1344 (2010)

25. Soeken, M., Wille, R., Drechsler, R.: Verifying dynamic aspects of UML models. In: Design, Automation and Test in Europe, Grenoble, pp. 1077–1082 (2011)
26. Soeken, M., Wille, R., Drechsler, R.: Assisted behavior driven development using natural language processing. In: International Conference on Objects, Models, Components, Patterns, Prague (2012)
27. Soeken, M., Wille, R., Drechsler, R.: Eliminating invariants in UML/OCL models. In: Design, Automation and Test in Europe, Dresden, pp. 1142–1145 (2012)
28. Warmer, J., Kleppe, A.: The Object Constraint Language: Precise Modeling with UML. Addison-Wesley Longman, Boston (1999)
29. Weilkiens, T.: Systems Engineering with SysML/UML: Modeling, Analysis, Design. Morgan Kaufmann, San Francisco (2008)
30. Wille, R., Soeken, M., Drechsler, R.: Debugging of inconsistent UML/OCL models. In: Design, Automation and Test in Europe, Dresden, pp. 1078–1083 (2012)

Chapter 4
TLM POWER3:
Power Estimation Methodology for SystemC TLM 2.0

David Greaves and Mehboob Yasin

Abstract We report on a SystemC add-on library which extends every SystemC module with non-functional data regarding power consumption and physical layout and which accumulates and estimates dynamic energy usage. It supports both phase/mode power modelling and energy-per-transaction logging for TLM (transactional-level modelling). Wiring energy is computed by counting bit-level activity within the TLM generic payload. Each leaf component can also register its physical dimensions to facilitate a wire length estimator that traverses the SystemC model hierarchy using either full placement or Rent's rule estimators. It also supports dynamic voltage islands and inter-chip wiring, where each transaction can consume energy according to the current supply voltage of the relevant islands and the nature of the interconnect. We report on basic performance from some SPLASH-2 benchmarks running on a modelled OpenRISC quad-core platform.

4.1 Introduction

With the current major emphasis on power consumption in electronic design it is important to be able to obtain power estimates during the architectural exploration phase. Power consumption is an emergent property arising once hardware and software have been selected. For results to be numerically accurate, a detailed, net-level layout of the design is required in the chosen target technology. This level of detail is inconsistent with rapid prototyping. However, with wiring power becoming

---

D. Greaves (✉)
Computer Laboratory, University of Cambridge, Cambridge CB3 0FD, UK
e-mail: David.Greaves@cl.cam.ac.uk

M. Yasin
Computer Laboratory, King Faisal University, Al-Ahasa, Saudi Arabia
e-mail: my294@cl.cam.ac.uk

J. Haase (ed.), *Models, Methods, and Tools for Complex Chip Design*, Lecture Notes in Electrical Engineering 265, DOI 10.1007/978-3-319-01418-0_4,
© Springer International Publishing Switzerland 2014

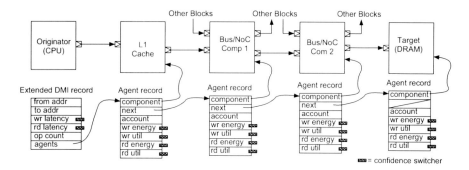

Fig. 4.1 An example extended DMI record and agent list. An initiator may typically have several of these active at once for different targets or addressable regions in a target. The TLM return path is always the same as the forward path and the agent records are incremented for utilisation and energy in (the active phase of) both directions

the dominant contributor in recent generations of VLSI technology, early indications of this aspect are becoming more essential. Indicators that are relatively accurate become useful. Relatively accurate indicators may have unknown linear error factors in the values they report, but they certainly have the correct polarity in their partial derivatives, thereby allowing the designer to tell whether a change is better or worse.

TLM modelling using SystemC permits high-performance models to be created. The greatest performance is facilitated by using the blocking transaction style with loose timing (L/T) and DMI (direct memory interaction). Using blocking transactions, interactions between a CPU and a cache, memory or I/O device are modelled as a simple method invocation with handshaking overheads being modelled simply by the call and return of the relevant subroutine [3]. The loose timing method allows a given initiator to hog the modelling workstation for an extended period of time, called its quantum, and thereby avoid the overhead of context switching needed to keep transactions and bus cycles strictly in the order they would really occur. DMI allows an initiator, such as a CPU, to make backdoor access to the workstation memory used to model the contents of RAMs and DRAMs, thereby avoiding the overheads of modelling caches and busses or NoCs (networks on chip). However, previous modelling systems have become highly inaccurate in terms of reported performance and (especially) power when these advanced modelling features are enabled.

Two previous libraries for SystemC power modelling are TLM POWER2 [6] and PKtool [13]. Our own library is called TLM POWER3 owing to its direct reuse of some infrastructure from TLM POWER2, but ideally one might merge it with PKtool so that the styles and approaches from both previous libraries are concurrently available. Higher-level approaches might also be included. For example, the Sesame approach to estimating power consumption uses an abstract model of execution, based on *computational event signatures* [8]. A similar higher-level approach was presented in [10], but built on SystemC.

4 TLM POWER3: Power Estimation Methodology for SystemC TLM 2.0

The TLM POWER2 library for SystemC associated groups of SystemC modules with a pair of power account records called static and dynamic. The association was maintained either by inheriting a `pw_module` parent as well as the standard `sc_module` inheritance, or by setting a SystemC attribute to point to the appropriate set of accounts.[1] TLM POWER2 used the mode/phase approach to power modelling.

In the mode/phase approach, the consumption of an IP block is determined from its current state. The state of the block is characterized by both its phase and its mode. A mode is a particular DPM (Dynamic Power Management) mode (e.g. on, sleep, off). A phase, basically a functional phase, is characterized by its power and time duration (e.g. wait, read, compute). The available modes and phases are defined in a technology/instance file that is inspected by the constructor for the component. The constructor can nominate a specific file for a specific instance or the `kind()` of the component can be looked up and the details set for all instances of that kind of component. The behavioural model for the block must change mode and phase explicitly using calls such as

```
this->update_power(sc_pwr::PW_MODE_ON,
                   sc_pwr::PW_PHASE_IDLE);
```

Infact, in the mode/phase approach of TLM POWER2, there is no specific support for transactional modelling or loose timing. The TLM calls are unannotated and the SystemC kernel must be advanced for the appropriate period of time while a component is in a given power mode/phase for the correct energy accumulation to be logged.

PKtool is another SystemC library for power modelling, but its basic approach is to count transitions at the net level. Wrappers are provided for all of the common SystemC datatypes used for modelling wires, such as `sc_uint<7>`. When SystemC kernel time advances, the hamming distance of each wrapped type is computed and added to its transition count. The hamming distance is the number of bits that have changed value. For energy modelling, only the zero-to-one transition needs to be considered. A net will consume energy from the supply each time it rises, according to the standard $\frac{1}{2}CV^2$ formula, where C is the net capacitance, which is proportional to its length. (As explained below, we use the same approach for our TLM calls, but we then automatically disable it in favour of performance). However, PKtool library does not help estimate net length, and despite some recent extensions for TLM modelling, it has no support for the TLM generic payload. Directly relating the events in the model to the SystemC kernel timestamp cannot support loosely-timed models which locally run ahead of the kernel.

---

[1] We use the word *component* to denote an `sc_module` that is so associated.) SystemC augments every `sc_module` (or other entity that inherits `sc_object`) with a key/value space where the values are `void *` pointers.

4.2 Our Approach: TLM POWER3

TLM POWER2 defined physical units for power and energy in the same way as SystemC itself defines physical units for time. All of the standard arithmetic operators are overloaded to have the expected behaviour. For instance, a power multiplied by a time results in an energy. In TLM POWER3, we have added new physical units for voltage, distance and area, along with the appropriately overloaded operators. A component can describe its physical size in its constructor using one of the following TLM_POWER3 calls:

```
// Set actual dimensions of current component
void set_fixed_dimensions(pw_length x, pw_length y);

// Set additional area of current component
void set_excess_area(pw_area a, float max_aspect_ratio=2.5);

// Select chip/voltage island for current component
//   and its children.
void set_chip_name(string chipname, string island);
```

The former sets the actual dimensions of the current component, leading to a warning if this is smaller than the sum of its components. The excess_area call describes the additional area of the current component beyond that of the sum of its child components. The component is assumed to be flexible in shape from square up to an oblong of maximum aspect ratio specified. Aspect ratio is, however, ignored by our provided basic estimator that just sums areas within a component and does not attempt to give them co-ordinates within the component. Components can be specified to be placed on different chips or regions of chips but the default is to be on the same chip/region as their parent. This identifies which wiring crosses between chips and hence has different dimensions and technology. It can also be used to exclude logic from the current chip's dimensions, as is useful for example, when a DRAM bank model is instantiated inside the DRAM controller rather than exporting all of the connections (TLM or otherwise) up through the module hierarchy. The same partitioning approach defines dynamic supply voltage islands where voltage changes are applied to all members of a chip/region at once.

As well as supporting an external table of modes and phases for each instance/kind of component we enable the C model to contain explicit statements of power and energy. For instance, the constructor (or PVT callback, mentioned later) for an SRAM of m_bits might contain the following, where the first line creates a constant power value and the second logs this power in the static power account of the current component.

```
pw_power leakage = pw_power(82.0 * m_bits, PW_nW);
set_static_power(leakage);
```

Rather than just supporting a fixed pair of power accounts, as in TLM POWER2, our library supports any number of accounts per group of components with the first three being nominally used for component static power, component dynamic power/energy and wiring dynamic energy. For full flexibility, each account can model both energy and power. Each account has energy as its primary accumulating representation and the power being a standing value that, form time to time, is converted to energy debits. Standing power is converted to energy when the standing power level is changed or at the end of simulation, the energy being the previous standing power level multiplied by the time since the last standing power change. An error is raised if the simulation exits at infinite time with a non-zero standing power level in any account. Energy figures are converted back to average power in some forms of report.

Our library also supports utilisation and transaction logging for visualisation purposes. Although this might seem orthogonal, there are some overlaps. One feature of the PKtool TLM modelling style is that idle power in a component is not accumulated while a transaction is active, and hence details of component utilisation are needed for this style of modelling. By recording utilisation we can apply this correction if desired: it might be very useful to model dynamic power and clock gating. In addition, when our library generates a VCD (Verilog change dump) report, it is convenient to have a graphical illustration of the transactions alongside the energy use bumps.

4.2.1 Extended Generic Payload: Distance + Hamming

Although our library can be used with the standard generic payload, more detail is captured using our extended version called PW_TLM_PAYTYPE. In TLM 2.0, sockets are templated types which default to use the standard generic payload, but we can instead use PW_TLM_PAYTYPE. Rather than explicitly extending the generic payload, we could have used the generic payload's own extensions (and these still work, as used for instance for extended commands such as load-linked/store conditional), but we chose not to for efficiency reasons. Socket definitions and calls now look like this (although CPP macros can tidy this up):

```
// Providing the third template argument to a socket:
tlm_utils::simple_initiator_socket
<mytype, 64, PW_TLM_TYPES> ifetch_socket, data_socket;

// Using the extended payload in the callbacks:
void b_access(PW_TLM_PAYTYPE &trans, sc_time &delay)
```

The extensions in PW_TLM_PAYTYPE assist with the following details:

1. Deciding which fields are active so that only the correct fields have their hamming distance processed for wiring power,
2. Establishing the trajectory of the transaction through the system so that traversed wire length is estimated,

3. Keeping note of the components encountered so that the correct power and utilisation accounts can be incremented under DMI,
4. Measuring the variance of metrics so that automated transition to DMI is enabled.

In a generic example, Fig. 4.1, the originator (CPU) will complete the address field of the payload and, for writes, also the data and byte-enabled fields. Generally, `multi_passthrough` TLM sockets are used in complex system models: these support forwarding the transaction onwards through bus, cache and NoC (network-on-chip) subsystems. The return path is always the reverse of the forward path owing to simple stack unwinding associated with method invocation. So intermediate components forward the payload, perhaps with minor changes (e.g. address space manipulations at bus bridges or VM units) to the target destination. This target will reply with a low-cost acknowledgement for a write and with the data for a read.

Our TLM payload offers an API with three library calls for bus energy modelling. These are `pw_set_origin`, `pw_log_hop` and `pw_terminus`. Currently models invoke these on a payload at the beginning, intermediate steps and end of a its payload trajectory. Rather than manual application, building these invocations into the TLM convenience sockets would be more convenient (sic).

```
void pw_set_origin(sc_module *where,
    uint flags=0,
    bit_transition_tracker *transition_reference=0);
pw_agent pw_log_hop(sc_module *where,
    uint flags=0,
    bit_transition_tracker *transition_reference=0);
void pw_terminus(sc_module *where);
```

The first argument `where` is the `this` pointer for the current component. This is used to track the path through the system.

The second argument is the flags that denote which fields in the payload are active. They can also encode bidirectional data busses and multiplexed address-data busses. When the physical nets of busses are re-used the transition count increases but there are fewer busses (e.g. the high order address bits might be mostly static on a dedicated address bus but are not if the same wires carry multiplexed address and data). Most flags are sticky and apply to subsequent hops that do not change that flag. In particular, if the flags argument is zero for the next hop then nothing has changed and the next hop has the same properties as the previous hop.

The third argument is a bus reference. Every transaction is considered to take place over a bus and a bus is a generic set of wires modelled with a `bit_transition_tracker`. Wires present (i.e. payload fields) that are not used consume no energy, so it is not important to customise the instance of a bus to its use (e.g. the bus from CPU to memory has address and write data whereas the return bus has just read data). The bus reference is needed so as to check which physical nets are transitioning with respect to their previous value.

We could integrate a layout package to estimate wiring lengths in detail. Currently we use the Rentian approach of [4] that provides a simple estimate

4 TLM POWER3: Power Estimation Methodology for SystemC TLM 2.0   59

of average connection length in a well-placed implementation according to $\alpha.A^{\frac{1}{2}}$ where A is the area of the lowest common parent component to the source and destination of the signal and α is typically 0.3. The capacitance per unit length of nets for on-chip and off-chip wiring is read from a configuration file (we use 0.3 pF/mm).

TLM 2.0 defined a DMI record called `tlm::tlm_dmi` which stores the start and end addresses and access times for a region of memory that can be accessed via a fast backdoor mechanism: the client simply reads from the raw host memory that contains the memory contents. In addition, the target has a callback called `invalidate_direct_mem_ptr` whereby this record can be retracted. However, using this DMI mechanism bypasses the target model and also all intermediate bus components, including caches, so their utilisation and energy accounts cannot be updated directly during DMI accesses. In TLM POWER3, we make the forward trajectory of a TLM DMI-allowed call instantiate a chain of account records that contain the energy and utilisation and account number for each intermediate agent and the target. The energy and utilisation are also updated on the return transit of the thread. (For non-blocking calls, the updates are just made on the significant protocol phases.) At the initiator we augment the DMI record with a count of the number of DMI calls made (scaled by the relative size of the transaction if the payload burst size is varying). Aperiodically (e.g. at end of L/T quantum), and on DMI invalidate, the count field is reset with the appropriate credits being made to the utilisation and energy accounts of each referenced component. The invalidate DMI callback also performs such a flush and frees the agent list. Operations such as store conditionals must not use DMI, so can be used as flush points.

An alternative to building the agent records is to write DMI energy to a 'slush fund' account where it will appear (correctly) in the total for the system/subsystem but (incorrectly) in the slush fund of the originator of such transactions instead of the consuming component (which is ensured by our agent records). We would perhaps use account number 4 in each component for this purpose.

The energy and power figures in a call to the library can either be pre or post supply voltage scaling, where the former are multiplied by the supply voltage squared at the point of logging and the former are not scaled. Given that a component (`sc_module`) inherits our `pw_module_base`, transaction energy logging in a component can be as simple as:

```
m_read_energy_op = pw_energy(5.0 +
    1.5e-5 * m_bits, pw_energy_unit::PW_pJ);
m_read_energy_op *= get_current_vcc_squared();
this->record_energy_use(m_read_energy_op /*, 1*/);
this->record_utilisation(sc_time(1, sc_us), delay);
```

where the first line would typically be in the component's constructor, the second would be in the constructor or in the PVT (process/voltage/temperature) recalculate callback. The third line actually logs the energy and can specify an alternative account to the default of '1'. Multiplying by the supply voltage squared on every logging event is slow, and hence pre-computing this in the PVT method is preferred.

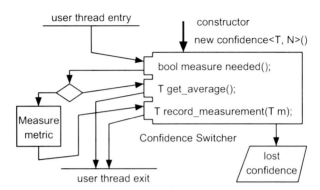

Fig. 4.2 General use pattern for the 'confidence switcher' component that first accumulates and then provides a mean value for a metric based on aperiodic measurements while raising an exception if accuracy is lost

Account one is the default intra-component dynamic energy account. The log of the utilisation itself takes the busy duration and an extra, optional second argument which is the advance over kernel simulation time needed for accurate rendering when loosely-timed. The 'this->' prefix would either be missed out, but preferable is is to replace it with the agent handle returned from log_hop call. This applies the energy and utilisation debits to the current component but also inserts their values in the agent list (if one is being constructed) so that they are accounted when subsequent calls are replaced with DMI.

Using standard TLM 2.0, an initiator will start using DMI when calling get_direct_mem_ptr on the initiating socket after a transaction instantiates a valid, local DMI record. Typically the initiator has no knowledge of the accuracy of the latency figures in its DMI record: naively, these will just be the result of the first call (which could be much slower owing to cache misses etc.). We provide and use a '*confidence switcher*' to ensure DMI is employed with fairly accurate values for latency as well as energy and utilisation in an extended DMI record.

The confidence switcher (Fig. 4.2) is a simple library component designed to capture the value of a presumed-stationary statistic using a relatively small number of costly trials. It has internal state and three user methods and is parametrised by an integer N (default is 1,000) that averages a generic statistic over the second N measurements and then reports that average from then onwards while making pseudo-random occasional further measurements (with mean spacing every $1/N$) to check that the mean value has not significantly changed. The first N measurements are not included in the average to avoid start-up transients. This gives a performance boost of approximately N times in the overhead of this measurement. A change by more than 1% and more than $2/N$ is considered significant and this raises an SC_ERROR or SC_WARNING according to another construction parameter. Confidence switchers are used as much as possible, both in the POWER3 library and by the user models. They can record bit transition density counts, latency times and and power and energy units.

```
#####################################################################
#                    TLM POWER3 (Univ Cambridge, UK)               #
#                                                                   #
#           Statistics file: energy/power consumption.             #
# ----------------------------------------------------------------- #
# For more information see the TLM POWER3 manual pdf.          p    #
# ----------------------------------------------------------------- #
# Creation Date: 17:27:22 -- 15/09/2012                             #
#####################################################################


Title: privmem-c1n6000-dramsim-withcache-nile-gash-harvard
# Simulation duration: 24826590001096 ps
# Simulation duration: 24826590001096 ps

+----------------------------+-------------------------+-------------------------+-------------------------+
| MODULE  NAME               |       STATIC0 ENERGY |       DYNAMIC1 ENERGY |        WIRING2 ENERGY |
+----------------------------+-------------------------+-------------------------+-------------------------+
Standalone modules:
| Memory 0 (DRAM)            | 0.173879501J   3.49% | 0.0875462788J   1.76% | 4.48687512e-07J   0.00% |
|  the_top.uart0            |          0J   0.00% |     1.644e-06J   0.00% | 6.7041e-11J   0.00% |
|  the_top.busmux0          |          0J   0.00% | 1.1905216e-05J   0.00% |          0J   0.00% |
|  the_top.dram=0           | 0.173879501J   3.49% | 0.0875462788J   1.76% | 4.48687512e-07J   0.00% |
| ...top.coreunit_0.core_0  | 0.2482659J   4.99% |  0.00440126263   0.09% | 1.34648772e-05J   0.00% |
| ...reunit_0.l1_d_cache_0  |          0J   0.00% | 0.000594064671J   0.01% | 6.14810556e-06J   0.00% |
| ...0.l1_d_cache_0.Data_0  | 0.0333542257J   0.67% | 0.000107935695J   0.00% |          0J   0.00% |
| ...0.l1_d_cache_0.Tags_0  | 0.0317907464J   0.64% | 4.18042825e-05J   0.00% |          0J   0.00% |
| ...0.l1_d_cache_0.Data_1  | 0.0333542257J   0.67% | 0.000105833853J   0.00% |          0J   0.00% |
| ...0.l1_d_cache_0.Tags_1  | 0.0317907464J   0.64% | 3.37903219e-05J   0.00% |          0J   0.00% |
| ...0.l1_d_cache_0.Data_2  | 0.0333542257J   0.67% | 0.000105435493J   0.00% |          0J   0.00% |
| ...0.l1_d_cache_0.Tags_2  | 0.0317907464J   0.64% | 2.60627187e-05J   0.00% |          0J   0.00% |
| ...0.l1_d_cache_0.Data_3  | 0.0333542257J   0.67% | 0.000108887529J   0.00% |          0J   0.00% |
| ...0.l1_d_cache_0.Tags_3  | 0.0317907464J   0.64% | 1.83743234e-05J   0.00% |          0J   0.00% |
+----------------------------+-------------------------+-------------------------+-------------------------+
```

Fig. 4.3 Part of an example textual report file where a large number of separate components have been included. The grand total account is clipped off the *bottom* of this figure

4.2.2 Output Reports

Several kinds of report can be generated with the TLM_POWER3 library. These cover power consumption, utilisation and physical layout.

The library will automatically add up the power and energy used globally, but further detail on individual sub-systems can also be reported by selecting other points in the hierarchy to trace and passing the associated component as an argument to a power trace function. Each item traced generates either a fresh set of accounts that include that item and all of its children, or that item alone, or a fresh set of totals for each component (i.e. for that item alone and recursively for all its children separately).

Energy consumption and average power are primarily reported in a plain text file emitted at the end of simulation (or at other user request dump points). An example is shown in Fig. 4.3. This file can also be written in spreadsheet-friendly SYLK (SYmbolic Link) form. As mentioned, utilisation, energy and transaction activity reports are available in VCD form. The VCD generator can also output in a gnuplot-friendly multi-column file format. Figure 4.4 shows an example VCD plot. The L/T (loosely-timed) approach can upset the normal SystemC VCD report format owing to temporal decoupling (events are logged in their actual simulation order rather than their nominal correct order) but our VCD writer corrects this by writing the events to a circular RAM buffer whose temporal extent is greater than the L/T quantum. This also enables sensible energy plots to be made: energy events would be like Dirac pulses if rendered directly and cumulative energy plots are not especially informative, but our VCD writer implements a single-pole low-pass filter for the

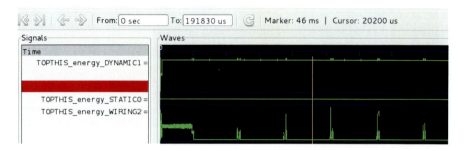

Fig. 4.4 An example VCD trace showing static, dynamic and wiring power consumption for a RADIX sort. The cores operate mainly from cache but exchange data between rounds

energy events so that they appear like exponentially-decaying pulses. Alternatively, in another mode, it reports the flat average power given by the last energy quantum divided by the time to the next-logged quantum on that account. Physical layout is currently printed as a text file which just reports which components are inside which others and the resulting area for each component. A graphical plot in .svg form is being implemented.

4.3 Performance

We examined the performance of the first two testbench programs in the Splash-2 suite [14]. These are a radix sort and a L/U matrix decomposition that can run on 1–16 cores. We compiled the Splash-2 programs to run bare metal supported by the standard linux libc and our own implementation of pthreads and a Simics/ANL (parmacs) shim layer [5]. Our testbench uses four OpenRISC processor cores [11] in verilated or fast ISS forms wrapped to use TLM 2.0 blocking calls served by an un-cached instance of DRAMSIM2 [9]. The cores log 250 pJ per instruction and run at 200 MHz unless paused waiting for other cores (50 mW core power). Each core has separate I and D L1 caches that included 17 RAMs each (tag and data for 8-way set associative and a write buffer). Each Core, Cache and each of the other components shown in Fig. 4.5 is a separately-annotated SystemC module that also inherits a TLM_POWER3 base and communication between them is completely with blocking TLM 2.0 calls. There are 16 SystemC modules in 3 levels of hierarchy. The individual RAMs had dimensions and power consumption computed according to the equations in Table 4.1 which were formed from our own regression of 45 nm CACTI runs [12].

Table 4.2 shows that simply taking the four-core ISS and putting it inside SystemC TLM degrades the performance by about ten-to-one owing to SystemC kernel overhead (gprof reveals major costs (more than 20% of execution time) are

4 TLM POWER3: Power Estimation Methodology for SystemC TLM 2.0

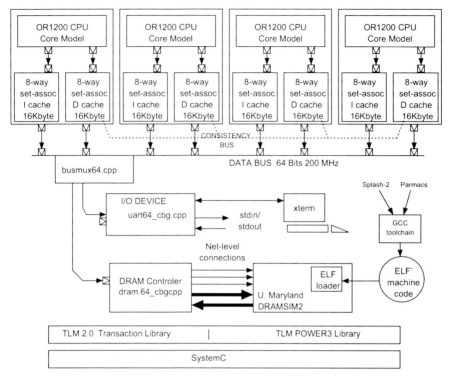

Fig. 4.5 Reference hardware platform for experiments (Quad-core OPENrisc with U. Maryland DRAM simulator)

Table 4.1 Interpolated CACTI 45 nm RAM parameter equations

Property	Model equation	Unit
Area	$13,359 + 40*bits$	squm
Read energy	$5 + 1.5E - 5$	pJ per bit
Write energy	$10 + 3.0E - 5$	pJ per bit
Access time	$0.21 + 3.E - 4 \times (bits)^{0.5}$	ns
Leakage power	82	nW per bit

Table 4.2 Simulation performance using GCC 4.43 on Intel x86_64 3 GHz/6,000 BogoMips, 8 GB RAM (no paging) SystemC-2.2.0

Figure 4.6 name	Configuration	Instructions/s	Ratio
Not plotted	Fast ISS – No SystemC	11×10^6	1.0
Unannotated model	L/T = min, POWER3 = off	1×10^6	0.1
Unannotated model	L/T = max, POWER3 = off	4×10^6	0.4
DMI enabled	L/T = max, POWER3 = off	7×10^6	0.7
POWER3	L/T = max, POWER3 = on	0.5×10^6	0.05
POWER3 + HOPS	L/T = max, POWER3 = on + hops	0.3×10^6	0.03
POWER3 + HOPS+XITIONS	L/T = max, POWER3 = on + hops + hamming	0.2×10^6	0.02

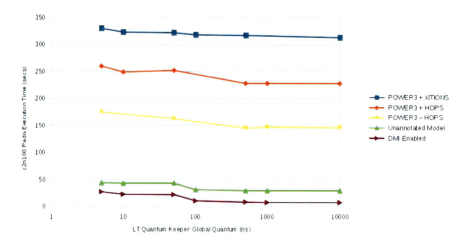

Fig. 4.6 Relative simulation performance of approximately-timed (*left-hand side*) and loose-timed (*right-hand side*) TLM Model (2 cores, running SPLASH-2 Radix Sort n = 100) with various configurations

in sc_core::sc_simcontext::crunch(bool) and b_transport. This degradation occurs with and without the inclusion of caches but the performance of the modelled system then changes as expected (i.e. program completes much faster with caches). The next lines in Table 4.2 are taken from Fig. 4.6 which plots the performance with and without DMI with respect to the L/T quantum keeper value. Using DMI and a maximal L/T quantum, so that the SystemC kernel is only entered during bus and mutex contention, restores some of the performance.

The effect of compiling with our power library with various configurations is reported at the bottom three lines of the table (and in further plots). It gives roughly a factor of two slow down and the logging of hops does not make it much worse (penultimate line). Implementing hamming distance computations under control of confidence switchers where with $N = 1,000$ causes a further rough factor of 2 performance degradation. Performance can perhaps be improved upon, but it is not overly bad. Interestingly, the degradation was much worse in an early version where the island voltage was squared at every use rather than recomputing the transaction energies just on each PVT change.

Figure 4.7 shows the measured power consumption of the processor (excluding DRAM) on a real Linux workstation as three identical runs of the RADIX benchmark were executed, the third one using only one CPU core. A 0.05 Ω resistor was put in series with the 12 V supply to the processor and its voltage drop and output voltage were logged at 60 Hz to record the energy consumption. Figure 4.8 shows the total power plot when the same C program was run on the SystemC model. Some differences in general shape are obvious and need investigation.

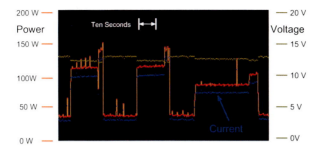

Fig. 4.7 Splash RADIX benchmark: probed processor power consumption. Two runs using two cores followed by one run on a single core. The end region of each run is the checking phase. Spikes are other unix processes on the dual-core workstation (Intel Pentium D 3 GHz)

Fig. 4.8 Splash RADIX benchmark: TLM POWER3 total power consumption: we see one run using two cores followed by one run using a single core. No unix operating system was present

4.4 Accuracy

To explore the general accuracy of our library we used three simple test programs to generate calibration data and used this data to predict results for a fourth program. Each of the four programs could be run with one to six threads. We measured the CPU energy use and execution time on a 2.4 GHz AMD Phenom X6 1045T six-core chip, as plotted in Figs. 4.9 and 4.10 respectively. This chip has 64 KB I + D caches per core as well as a dedicated 512 MB L2 cache per core and a shared 6 MB L3 cache. The tests could all run within the L3 cache so DRAM power did not need to be included. The test programs were respectively a memory-bound program with disjoint regions that each fit within the dedicated caches of a core, a memory-bound program with a moderate amount of inter-core churn and a CPU-bound program. Using a multivariate regression spreadsheet within Libre Office the coefficients in Table 4.3 were determined. These were then used to calibrate the SystemC models using the POWER3 library to predict the power consumption and performance of other programs. Such a program was the SPLASH-2 RADIX benchmark, run with between one and six cores and plotted as the final six results.

In each test the total amount of work was increased linearly with the number of cores. The energy used can be seen to grow roughly linearly as well but the execution time only grows when the cores contend for cache lines. The final six results show good agreement between measured and predicted values (within 30%) which is acceptable for high-level architectural exploration. We cannot expect perfect prediction since the programs were compiled for OpenRISC during simulation and

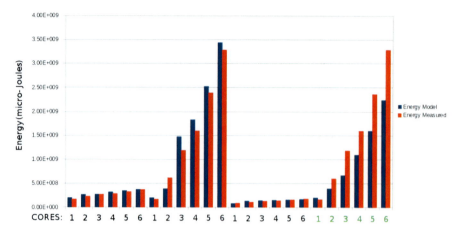

Fig. 4.9 Measured and modelled energy consumption for four tests each with one to six cores where coefficients from the first three tests were used to predict energy in the fourth

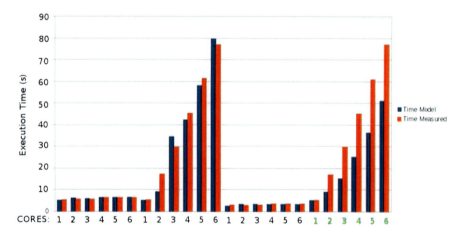

Fig. 4.10 Measured and modelled execution time for four tests each with one to six cores where coefficients from the first three tests were used to predict execution time for the fourth

Table 4.3 Energy debits obtained from curve fitting between simulation and measurement over 24 runs with 1–6 cores in use: CPU & Caches only (DRAM excluded)

Operation	Energy
Instruction	1 nJ
L1 + L2 I cache miss	50 nJ
L1 + L2 D cache miss	15 uJ
L2 cache snoop read	4 mJ
L2 cache consistency evict	7 mJ

for x86-64 during measurement. We have recently implemented a more-detailed model of AMD's hyper-transport system and Hammer cache protocols and will report more comprehensively in another paper.

4.5 Conclusion

Our framework provides an easy-to-use power estimation add on to SystemC TLM modelling. The use of the confidence switcher to dynamically disable detailed modelling is novel. The user may easily alter the system structure in radical ways, changing cache size, bus layout and so on. Standard ELF binaries can be easily generated with GCC/binutils tool chain. We also have a MIPS64 SMP system based around the same components. Because wiring power is becoming a dominant aspect it is important to include it in rapid exploration tools.

The benefit of rapidly exploring design options using SystemC was advocated in [1], but having performance predictions without power predictions is no longer acceptable. A fairly-detailed TLM model with power annotation was constructed by [2] for a PowerPC-based SoC. The activity for individual test transactions was extracted from VCD files and entered into a database. This approach can be applied in our framework to generate the individual transaction energies. Power estimation is also being performed for AMS (analog and mixed signal) subsystems within the SystemC framework [7].

We plan to further refine our API and library and release it for download. Including the `log_hop` operations inside the convenience sockets would be sensible. Also, further support for power islands might be needed, but currently we can use our chip number concept with zero volt supply setting to disable static power in regions. Further work will be to integrate back-annotation flows from real layouts and compare these with the Rentian approach. We would also like to extract net-level activity from the Verilated models to gain additional insight and confidence.

Acknowledgements We thank Matthieu Moy for providing the TLM POWER2 base platform.

References

1. Benini, L., Bertozzi, D., Bogliolo, A., Menichelli, F., Olivieri, M.: Mparm: exploring the multi-processor SoC design space with SystemC. J. VLSI Signal Process. Syst. **41**, 169–182 (2005)
2. Dhanwada, N.: A power estimation methodology for SystemC transaction level models. In: Proceedings of CODES-ISSS, Jersey City, pp. 142–147 (2005)
3. Ghenassia, F.: Transaction-Level Modeling with SystemC: TLM Concepts and Applications for Embedded Systems. Springer, Secaucus (2006)
4. Greenfield, D., Moore, S.W.: Fractal communication in software data dependency graphs. In: SPAA'08: Proceedings of the Twentieth Annual Symposium on Parallelism in Algorithms and Architectures, Munich, pp. 116–118. ACM, New York (2008)

5. Magnusson, P.S., Christensson, M., Eskilson, J., Forsgren, D., Hallberg, G., Hogberg, J., Larsson, F., Moestedt, A., Werner, B.: Simics: a full system simulation platform. Computer **35**(2), 50–58 (2002)

6. Moy, M.: Mini power-aware TLM-platform. http://www-verimag.imag.fr/~moy/?Mini-Power-Aware-TLM-Platform (2010)

7. Pêcheux, F., El Abidine, K.Z., Greiner, A.: Early power estimation in heterogeneous designs using SoCLib and SystemC-AMS. In: Proceedings of the 20th International Conference on Integrated Circuit and System Design: Power and Timing Modeling, Optimization and Simulation, PATMOS'10, Grenoble, pp. 252–252. Springer, Berlin/Heidelberg (2011)

8. Piscitelli, R., Pimentel, A.D.: A signature-based power model for MPSoC on FPGA. VLSI Des. **2012**, 6:6–6:6 (2012)

9. Rosenfeld, P., Cooper-Balis, E., Jacob, B.: Dramsim2: a cycle accurate memory system simulator. Comput. Archit. Lett. **10**(1), 16–19 (2011)

10. Streubuhr, M., Rosales, R., Hasholzner, R., Haubelt, C., Teich, J.: ESL power and performance estimation for heterogeneous mpsocs using SystemC. In: Specification and Design Languages (FDL), 2011 Forum on, Oldenburg, pp. 1–8 (2011)

11. Tandon, J.: The openrisc processor: open hardware and linux. Linux J. **2011**(212) (2011)

12. Thoziyoor, S., Ahn, J.H., Monchiero, M., Brockman, J.B., Jouppi, N.P.: A comprehensive memory modeling tool and its application to the design and analysis of future memory hierarchies. In: Proceedings of the 35th Annual International Symposium on Computer Architecture, ISCA'08, Beijing, pp. 51–62. IEEE Computer Society, Washington (2008)

13. Vece, G.B., Conti, M.: Power estimation in embedded systems within a SystemC-based design context: the PKtool environment. In: Seventh Workshop on Intelligent Solutions in Embedded Systems, Ancona, pp. 179–184 (2009)

14. Woo, S.C., Ohara, M., Torrie, E., Singh, J.P., Gupta, A.: The splash-2 programs: characterization and methodological considerations. SIGARCH Comput. Archit. News **23**, 24–36 (1995)

Chapter 5
SCandal: SystemC Analysis for Nondeterminism Anomalies

Jan Henrik Weinstock, Christoph Schumacher, Rainer Leupers, and Gerd Ascheid

Abstract SystemC is the de facto standard language for electronic system level design and simulation. SystemC simulations may contain nondeterminism caused by dependencies on the process execution order (PEO) due to data dependencies of SystemC logical processes (LP) within delta-cycles. In practice, often this is not an issue, since simulation execution *appears* to be deterministic due to deterministic SystemC scheduler implementations.

However, to satisfy the increasing need for simulation speed, parallel SystemC engines are being researched: With no fixed strict total order among LPs executed in parallel, nondeterministic behavior is more likely to surface and more difficult to debug, threatening the viability to use simulation for debugging use-cases.

This work presents a new method to test for nondeterminism: Anomalies are detected by running a simulation twice in sequential simulation mode while systematically varying the PEO, and without the need for source code analysis. Feasibility is demonstrated with several case studies.

5.1 Introduction

Embedded system hardware and software are becoming more complex, mainly due to the advent of multi-core technologies. Accordingly, system design complexity is increasing. Simulation techniques are employed to mitigate the risk of misdesigning such systems. Simulators provide system architects with early feedback about system behavior and performance before the actual hardware is completed.

J.H. Weinstock (✉) • C. Schumacher • R. Leupers • G. Ascheid
Institute for Communication Technologies and Embedded Systems, RWTH Aachen University, Aachen, Germany,
e-mail: jan.weinstock@ice.rwth-aachen.de; christoph.schumacher@ice.rwth-aachen.de; rainer.leupers@ice.rwth-aachen.de; gerd.ascheid@ice.rwth-aachen.de

J. Haase (ed.), *Models, Methods, and Tools for Complex Chip Design*, Lecture Notes in Electrical Engineering 265, DOI 10.1007/978-3-319-01418-0_5,
© Springer International Publishing Switzerland 2014

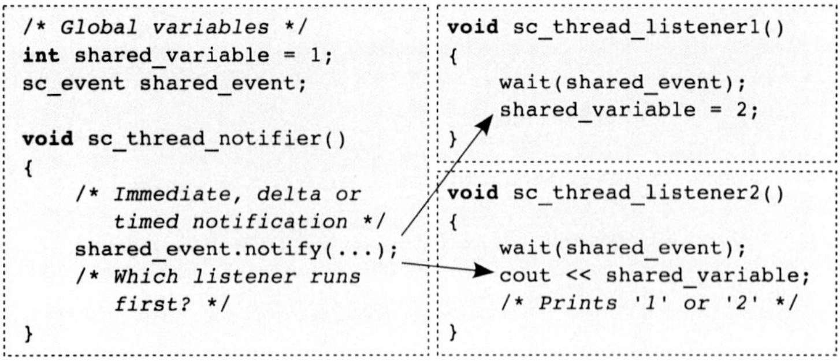

```
/* Global variables */              void sc_thread_listener1()
int shared_variable = 1;            {
sc_event shared_event;                  wait(shared_event);
                                        shared_variable = 2;
void sc_thread_notifier()           }
{
    /* Immediate, delta or          void sc_thread_listener2()
        timed notification */       {
    shared_event.notify(...);           wait(shared_event);
    /* Which listener runs              cout << shared_variable;
        first? */                       /* Prints '1' or '2' */
}                                   }
```

Fig. 5.1 Nondeterministic simulator example

Additionally, simulators offer to developers inspection capabilities beyond those of hardware for investigating software malfunctions ("bugs").

SystemC [11] is a popular C++ library to create simulators and executable specifications of embedded systems. It allows the modeler to utilize the full expressive power of C++ and provides all constructs commonly found in discrete event simulation (DES) systems to describe LP behavior.

Along with complexity, the gap between real-time and simulation time is also increasing. Two major orthogonal techniques exist to speed up simulations. Abstraction techniques like transaction level modeling (TLM) [18] reduce the number of simulated details and bypass the simulation engine for LP communication. Complementary, parallel simulation engines, e.g., [2, 15, 17, 20] spread the simulation over multiple simulator host or GPU cores.

Common model source code as of today is typically written with the sequential execution model in mind. A single LP activation is assumed to run uninterruptedly and exclusively. For parallel simulation, under certain circumstances mutex locks must be introduced if multiple LPs are accessing shared variables within the same delta-cycle. This can be done manually, or in some cases automatically [5, 14]. Another at least equally severe and intricate issue is that parallel SystemC simulation engines face the problem of establishing a strict total order between LPs as demanded by the SystemC standard [11]. This is especially true when faced with LPs communicating directly via shared variables bypassing the simulation engine, as it is often the case for TLM. Accordingly, without taking precautions, parallel simulation behavior may be nondeterministic. An example of a nondeterministic yet valid simulation is shown in Fig. 5.1: Here, it is undefined whether the output of the simulation will be "1" or "2".

Nondeterministic behavior puts the viability of parallel simulation as a tool for hardware and software development at risk. In essence, a nondeterministic simulation environment strips away the ability to go back in simulation time by re-running a simulation up to an earlier timestamp. This ability is essential for analyzing software malfunctions when race conditions are present in the simulated system.

In this paper we present *SCandal*, an extension to the OSCI SystemC 2.2 reference distribution [11] for detecting dependencies on the process execution order in simulations. The key idea is to compare observable behavior of a reference execution with that of a provocation run. In the latter, the order of all possible logical process pairs that are run is inverted, while respecting constraints arising from immediate notifications. While the original motivation for this research is to support the preparation of models for parallel simulation, the resulting tool is equally well suited for testing sequential simulations for well-defined, deterministic behavior.

The key contributions of this work are as follows:

Usable with binary model libraries. The proposed methodology does not require model source code analysis or annotation, nor the observation of communication between LPs. In contrast to previous works in the SystemC domain, *SCandal* can still do a systematic analysis of the system under test, without resorting to random testing as in [10].

Transfer of general debugging research to SystemC. *SCandal* focuses on situations that are likely to lead to bugs or other anomalies as identified in recent works such as [13, 16].

Simulator observation. Techniques to detect changes in simulation behavior are discussed, with focus on nonintrusiveness.

Anomaly detection. A set of new algorithms is presented that identify PEO dependency issues that are likely to appear: First, their presence is tested for by running a given simulation exactly twice. Succeeding stages then determine the time of causation and the involved LPs.

Prototype realization. Results of applying the proposed algorithms to a variety of well-known simulations, including all OSCI SystemC self tests and examples from model checking tools, the OSCI TLM 2.0 package and SoClib library, are discussed.

The remainder of this chapter is structured as follows: First, a brief overview of SystemC is given in Sect. 5.2. After that in Sect. 5.3, related work is introduced. Section 5.4 gives details on the nondeterminism anomaly detection algorithms as well as on the realization of the augmented SystemC kernel. Section 5.5 discusses experimental results, and finally, in Sect. 5.6 conclusion and outlook are presented.

5.2 SystemC Simulation Concept

SystemC is a *discrete event simulation* framework. Simulations consist of LPs that exchange information with each other via either *channels* or *plain C++ shared variables*. SystemC time is divided into two hierarchical levels: *timestamps* and *delta-cycles*. Delta-cycles add structure to timestamps and are typically used to conduct deterministic updates of shared variables stored in channels.

Events are *notified* by LPs, which need to specify the time at which an event will be *triggered*. When an event is triggered, LPs that are sensitive to this event become *runnable* and are put into the scheduler execution queue.

Event notifications can indicate three different types of time: First, a *timed* notification indicates the first delta-cycle of a future timestamp, second, a *delta* notification indicates the current timestamp but the next delta-cycle, or third, an *immediate* notification indicates the current timestamp and the current delta-cycle. Immediate notifications will receive special attention in later sections.

A possible scheduling algorithm is shown in Algorithm 2.

Algorithm 2: SystemC event loop (OSCI-based, simplified)

```
 1: while timed events to process exist do
 2:     trigger events at that time
 3:     while runnable processes exist do
 4:         while runnable processes exist do {In-delta loop}
 5:             execute one runnable process and trigger events
                receiving immediate notifications while executing
 6:         end while
 7:         update channels
 8:         trigger all delta time events
 9:     end while
10:     advance time to next event time
11: end while
```

5.3 Related Work

For the sequential case, [8] discusses various approaches to ensure that SystemC simulations behave deterministically independent of the scheduler implementation.

Research on parallel SystemC. and related environments rarely picks up the issue of nondeterministic behavior. Among those, Simics [14] ensures deterministic behavior by exploiting message delays and explicit communication channels, SPRINT [4] guarantees determinism by analyzing and transforming SystemC TLM simulations into Kahn process networks, and Parallel SimOS [19] deliberately settles with nondeterministic execution to attain higher simulation speed.

Detecting nondeterminism. is possible with a variety of approaches that are introduced below:

Model checking (MC) approaches have been successfully applied to SystemC simulations to prove design properties (e.g., [1, 3, 7]). MC provides the best flexibility, most comprehensive coverage and most extensive guarantees of all available methods. Typically, classical MC techniques operate on an abstract system description.

As the size of systems to check grows, the complete exploration of the model state space may become intractable. This is known as the state space explosion problem. Especially if models are generated from SystemC code, this is likely to happen, since static conservative C++ pointer analysis is often not able to sufficiently prune the model state space. Additionally, typically only a limited subset of SystemC or C++ functionality is supported, e.g., no *sc_spawn*, no *sc_next_trigger* etc. in [1], and no pointers, malloc, arrays, structs etc. in [3].

Dynamic partial order reduction (POR) approaches alleviate the state space explosion problem. A simulation is repeatedly executed, and decisions with indeterminate outcome are systematically varied to cover the state space completely or in part. In contrast to classical model checking, such approaches only require information about state changes and communication between LPs (e.g., [6, 9]). While typically less time consuming than MC, executing a given simulation possibly many thousand times and more may be infeasible.

Trace driven approaches typically need to execute a simulation only once for a given input, recording a set of events. This event trace is then analyzed to recognize patterns that could lead to assertion violations. Sen et al. [22] applies a trace-driven approach to detect race conditions and deadlocks, e.g., by analyzing the possibility of missing sc_event notifications. Trace driven approaches typically require source code instrumentation to be able to identify issues related to data shared directly between LPs.

Testing approaches are less common to analyze nondeterminism issues. In [10], scheduling of SystemC processes within a delta-cycle is randomized. Using assertions, simulations are tested by repeated execution without changing the input parameters. The tool can operate without source code instrumentation, but gives no guarantees which anomalies are detected.

Outside the SystemC world, CHESS [16] tries different sequentialized interleavings of the OS threads of multithreaded programs to find assertion violations.

5.4 Process Order Dependency Test

The basic idea of *SCandal*, a tool conducting SystemC Analysis for NonDeterminism AnomaLies, is to first execute a simulation with a reference process execution order while observing simulation behavior. In successive executions, the PEO is systematically varied while searching for deviations from the reference behavior.

In this section, the steps towards the design of the tool are outlined. First, mechanisms for behavior observation are introduced in Sect. 5.4.1. Afterwards, the class of targeted nondeterminism anomalies is defined in Sect. 5.4.2. Then, the required modifications to the OSCI SystemC scheduler implementation are discussed in Sect. 5.4.3. Finally, details of the analysis algorithms are presented in Sect. 5.4.4.

5.4.1 Behavior Observation

SCandal compares behavior of multiple simulation executions with different PEOs. Unexpected or missing observations are treated as nondeterminism anomalies.

An observation is a unique quadruple comprising the ID of the active SystemC process, the invocation count of this process in the current delta-cycle, the observation count of this process and the observation body. The available observation mechanisms focus on simulation output, heap activity, random number generation and LP activation. The user is responsible to select a feasible set of observation mechanisms.

5.4.1.1 Standard Output

Usually, SystemC simulators and C++ programs in general generate output via two different APIs: the Standard C Library and the C++ Standard Template Library. A mechanism to observe output therefore has to intercept calls to these APIs and record the string that is being sent to standard output.

Standard C Library Output. To capture output from the Standard C Library, calls to *printf* have been redirected by redefining the word *printf* using a preprocessor macro to *custom_printf*. The new function *custom_printf* is part of the observation mechanism and can be called in the same way as *printf* using a variable parameter list. Internally, the output string is recorded and then relayed to the standard output.

A non-invasive but non-portable approach would redirect calls using the linker. The GNU linker *ld* [26] is able to rename symbols during compilation. A strategy to replace all calls to *printf* with calls to *custom_printf* is to first rename *printf* to *real_printf* and then rename *custom_printf* to *printf*. The original *printf* is still available under the name *real_printf*.

The advantage of this method is that the code does not need to be recompiled, it must only be relinked. Therefore it is compatible with simulators that make use of precompiled third-party models.

C++ Library Output. The Standard Template Library offers the class *std::ostream* from which two instances *std::cout* and *std::cerr* are created upon program start. These are used to send data to the standard output stream and the standard error output stream, respectively.

Observation mechanisms must intercept calls made to those instances in order to capture output. This has been achieved by replacing the internal buffers that instances of type *std::ostream* operate on. These buffers have been derived from the originally used *std::streambuf* class and provide an overloaded *overflow* method that records any characters that did not fit into the buffer due to size limitations. To ensure that *overflow* is called for every character written to the buffer, the buffer size has been set to 0.

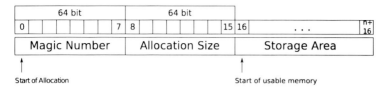

Fig. 5.2 Custom heap memory allocation layout

An alternate approach would replace *std::cout* and *std::cerr* with custom streams recording output using alternate implementations of the commonly used ≪ output operator. Unfortunately, those streams cannot easily be replaced, since their assignment operators (=) have been disabled. While there are methods available to bypass this limitation, for the sense of simplicity and to offer the best compatibility with third-party source code, the buffer replacement method as described above has been chosen instead.

5.4.1.2 Heap

The heap observation mechanism keeps track of the total number and size of memory allocations and deallocations made within a single delta-cycle by an LP on the heap. To achieve this, the default memory provider from the Standard C Library, i.e., *malloc* and *free*, have been replaced with a custom provider using a similar preprocessor macro as the one described in Sect. 5.4.1.1. Also, the *new* and *delete* operators that are usually used in C++ programs have been overloaded with new versions requesting and deleting memory using the same custom provider.

This custom memory provider offers two functions called *allocate* and *deallocate* that can be used for requesting or releasing heap memory. Besides allocating and deallocating heap memory, these functions keep track of the number and size of allocations and deallocations using a static global storage.

To be able to retrieve the size of an allocation upon releasing it, the memory provider stores this information with the allocation itself. This method was chosen over alternative methods like registries or maps, since these usually scale poorly with increased amounts of allocations.

Upon allocation, the memory provider asks the operating system for a heap region with 16 bytes more space than originally requested. The extra space is used to store a magic number as well as the size of the heap region. The effective layout of a single heap allocation produced with this mechanism is shown in Fig. 5.2.

When calling *deallocate*, the memory provider will check the magic number identifier of the allocation that should be freed. It retrieves this number by subtracting a constant offset from the allocation pointer, effectively looking at the memory in front of the allocation. If the magic number is incorrect, i.e., it has value different from the value that is assigned to all allocations by the *allocate* function, no memory size accounting can be done. In this case it must be assumed that the allocation is erroneous or was done by a different memory provider. If the heap identifier is correct, the size of the allocation is retrieved and accounted towards the deallocation size data of the currently active LP.

Once an LP has finished its execution for a given delta-cycle, the memory provider generates an observation including the heap statistics it has collected since the LP was started or woken up.

In some cases it might be beneficial to exclude allocations from certain sources. A typical example for this is a model that performs *lazy initialization*. All allocations that are done this way would always be counted towards the first process that accesses this model and thereby triggers the lazy initialization mechanism. Assuming multiple LPs access the model, it depends on the PEO which process actually performs the initialization. This is results in nondeterministic behavior, since the PEO is undefined.

Once it has become clear that a model behaves in such a way and it can be guaranteed that such behavior is not problematic, the lazy initialization can be excluded from observation. This is done using a *whitelist* that contains the C/C++ names of all functions that should not be accounted towards the heap statistics of an LP. Before the allocation is counted towards the heap statistics, the memory provider tries to identify the name of the function making the request. This is done by producing a call trace using the gcc specific functions *backtrace* and *backtrace_symbols*. This call trace is then checked if it contains a whitelisted function. If that is the case, the memory allocation will not be added towards the allocations of the currently running LP. Since this feature is computationally expensive, it is disabled by default and must be enabled manually by the user at compile-time using a preprocessor define.

5.4.1.3 LP Activations

The SystemC processes running during every delta-cycle are recorded. In case a process runs multiple times during a single delta-cycle, additionally the number of invocations is stored. If immediate notifications are used to activate an LP, the observation mechanism also stores which LP triggered the execution.

The behavior description generated with this observation mechanism serves two purposes. First, it is used to check if the processes are invoked within the correct delta-cycle. Second, the information is used as a reference PEO from which variations are computed systematically for later test iterations.

5.4.1.4 Random Number Generator (RNG)

Random number generators (RNG) are widely used in SystemC simulations, e.g., to simulate random traffic load or random input patterns for a system or component under test. This kind of intended non-determinism needs to be filtered out before looking for unintended nondeterminism like those caused by PEO dependencies.

During each simulation run, every LP must always see the same invariant sequence of random numbers as it saw during the initial execution.

A widely used RNG implementation is provided by the C standard library. It produces a sequence of pseudo random numbers given that it was initialized with a non-zero, random seed value. Usually the system real time clock is used for this purpose. Each time a simulation is run the real time clock will return a different value to be used as a seed, resulting in a different sequence of random numbers and possibly in a change of behavior of the simulator.

To counter this, the default RNG has been replaced with a custom one, allowing the user to specify a seed that will be reused for all test iterations to make sure random number sequences stay invariant at least for the duration of the test.

However, changes in the PEO might still result in LPs seeing different random numbers. Since the RNG hands out the random numbers on a first-come-first-serve basis, the individual sequence of numbers that an LP sees depends on its position in the execution order. Therefore it is necessary to provide independent RNGs for each of the LPs.

5.4.1.5 User-Provided

Previously described observation mechanisms focus exclusively on automatic observation of generic simulator behavior. They are not fit to describe anything specific about the behavior of the simulator since they lack the semantic context (e.g., a packet being relayed by a router model or the reception of an interrupt in a processor model). To increase simulator behavior coverage, the default observation mechanisms can be extended with user-provided observers.

5.4.2 Detectable Anomalies

Related research in the debugging and general purpose computing communities studies the patterns of common software malfunctions. Lu et al. [13] analyzed known software malfunctions of popular general purpose software packages: for the examined software, *96% of all encountered malfunctions could be triggered by establishing a particular order between just two sections of selected threads* of a multi-threaded program.

Additionally, [16] investigates how to expose bugs in general purpose software by sequentializing the execution of parallel programs and putting OS thread switching under the control of a modified scheduler. It is reported that typically *switching threads at one or two distinct points in time is enough to force a malfunction to appear.*

To summarize, a coverage that is viable in practice can be attained even by highly restricted search patterns. Transferring these results to SystemC, this work focuses the search on nondeterminism anomalies that satisfy the following assumptions:

- The anomaly can be observed (see Sect. 5.4.1).
- An anomaly is triggered by changing the execution order of exactly two LP invocations that are executed in the same delta-cycle.
- Changing the order of other process pairs does not prevent the anomaly from appearing.

The analysis framework is also able to find anomalies that do not satisfy these conditions, but their detection is not guaranteed (also see Sect. 5.4.4 below).

5.4.3 Controlled Scheduling

The testing algorithms described below require a number of enhancements over the standard SystemC scheduler:

Uniform treatment of sc_methods and sc_threads. The OSCI reference scheduler provides separate queues for both kinds of LPs that are always processed in succession. A generalized scheduling mechanism has been realized that supports mixed execution of sc_threads and sc_methods in arbitrary order.

Scheduling dependency tree (SDT). For every delta-cycle an SDT is stored recording which LP caused which LPs in which order to be scheduled by using immediate notifications (see Sect. 5.2). LPs runnable at delta-cycle start are recorded as children of the delta-cycle SDT root node.

5.4.4 PEO Dependency Analysis

The nondeterminism anomaly detection process comprises four stages, which are run in succession. Only the first two stages are mandatory, while the remaining stages try to derive additional information. The stages and applied algorithms are presented below; Table 5.1 summarizes the individual stage runtime complexities.

Stage 1: Reference Generation

The simulation to analyze is executed once. Enabled observers create the *reference observation database*. Moreover, SDTs (see Sect. 5.4.3) of every delta-cycle are recorded for use in later stages.

The key point during reference generation is that a newly proposed prepend-new-process-list (PNPL) LP scheduling algorithm is used inside the delta-cycle

5 SCandal: SystemC Analysis for Nondeterminism Anomalies

Table 5.1 Analysis stages runtime complexities

Stage		Req. sim. exec.	c_{end}
Reference generation		1	user set
Anomaly detection		1	user set
Causation delta-cycle detection		$\lceil log_2(c_{observation} + 1) \rceil$	$c_{observation}$
Dependant process identification	best	2	$c_{causation}$
	average	n_p	$c_{causation}$
	worst	$2n_p - 1$	$c_{causation}$

c_{end} Delta-cycle up to which the simulation is run
$c_{observation}$ Observation delta-cycle index
$c_{causation}$ Causation delta-cycle index
n_p Process activations in causation delta-cycle

scheduling loop. This allows to efficiently reverse the PEO later in stage 2. An algorithm close to the OSCI reference implementation scheduling and the PNPL scheduling algorithm are displayed in Algorithm 3; their difference can be seen in lines 6–10.

Algorithm 3: OSCI/PNPL delta-cycle scheduling algorithm

```
 1:  execution_list := list of runnable processes
 2:  scheduling_list := empty
 3:  while execution_list not empty do
 4:      extract and execute first process from execution_list
         {processes that became runnable via immediate-notifications
         are appended to scheduling_list}
 5:      if scheduling_list not empty then
 6:          if mode is OSCI then
 7:              append scheduling_list to execution_list
 8:          else if mode is PNPL then
 9:              prepend scheduling_list to execution_list
10:          end if
11:          scheduling_list := empty
12:      end if
13:  end while
```

Stage 2: Anomaly Detection

The goal of this stage is to decide whether a detectable anomaly (see Sect. 5.4.2) is present. To achieve this, the simulation is executed again, inverting the execution order of all possible LP pairs within delta-cycles while preserving causal relationships during a single simulator run. The algorithm to reverse the execution order is shown in Algorithm 4 and called causal-reversed-order scheduling (CRO) below.

Algorithm 4: CRO delta-cycle scheduling algorithm

1: reference_list := reference process execution order
2: execution_list := list of runnable processes
3: scheduling_list := empty
4: **while** execution_list not empty **do**
5:    intersect reference_list and execution_list
      {preserve order of reference list}
6:    select last process from intersection
7:    remove last occurrence of this process from both lists
8:    execute this process
      {processes that become runnable via immediate-notifications are appended to scheduling_list}
9:    **if** scheduling_list not empty **then**
10:      join scheduling_list to execution_list
11:      scheduling_list := empty
12:   **end if**
13: **end while**

While the simulation is re-executed with CRO scheduling, observations are compared to the reference observation database. Missing or unexpected observations indicate nondeterministic behavior.

In case the observations of reference and detection run match exactly, no detectable anomaly is present. Yet if still an anomaly not matching the criteria of Sect. 5.4.1 is suspected, a test mode that continuously re-executes the simulation with randomized PEOs is available to try to discover such anomalies by chance, similar to [10]. This mode is not used in the experiments presented in Sect. 5.5.

Reason to use PNPL and CRO. PNPL scheduling corresponds to a pre-order tree walk, and CRO to an anti-pre-order tree walk over the SDT (see Sect. 5.4.3). During a pre-order tree walk, for any given tree node P (see Fig. 5.3), all nodes P_L to the left of P or left to one of its ancestors will be visited before P; all nodes P_R will be visited in the same manner after P. For an anti-pre-order tree walk, the opposite is the case. For both kinds of tree walks, ancestors and descendants will always be visited before and after P, respectively. Accordingly, for any pair of nodes P and P_2 exactly one of the following two conditions is true: If P_2 is an ancestor or descendant of P, then their order will not change regardless of the tree walk being a pre-order or an anti-pre-order walk. Or, if P_2 lies inside P_L or P_R, then using an anti-pre-order tree walk instead of a pre-order tree walk will cause P and P_2 to be in reverse order.

Therefore, compared to PNPL scheduling, CRO scheduling changes the execution order of all possible LP pairs as long as causality is not violated.

Stage 3: Causation Delta-Cycle Detection

This stage determines in which delta-cycle an anomaly that was observed in stage 2 was actually caused. Therefore, a binary search is conducted as shown in Algorithm 5.

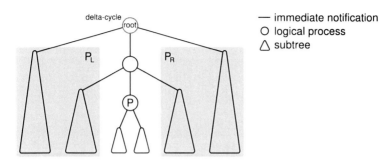

Fig. 5.3 Relationships in SDT

Algorithm 5: Binary search for causation delta-cycle

1: interval := [0 .. index of observation delta-cycle]
2: **while** interval size > 1 **do**
3:     execute simulation with PEO in first half inverted
4:     **if** anomaly appeared **then**
5:         interval := first half of interval
6:     **else**
7:         interval := second half of interval
8:     **end if**
9: **end while**
10: **return** index of single delta-cycle left in interval

To employ a binary search method, it is required that changing the PEO in multiple delta-cycles at the same time will not hide the anomaly. This is the case for all detected anomalies presented in the case studies (see Sect. 5.5). Regular linear search is required otherwise.

Stage 4: Dependent Processes Identification

The goal of this stage is to identify two LPs that, if their execution order is reversed, cause an anomaly to appear. The simulation is advanced to the causation delta-cycle $c_{causation}$ as detected in stage 3 using PNPL scheduling. The two methods for detecting two processes of a dependent LP pair inside $c_{causation}$ are described below.

First dependent process (stage 4a). In Fig. 5.4 an SDT is shown, with LP nodes arranged from left to right in the same order as they are executed when using PNPL scheduling. During successive simulation runs, each LP is individually moved to the back of the queue. Due to causal relationships, all descendant LPs must move together with the moved ancestor. Numbers indicate the order in which LPs are tested: Starting with processes executed last ensures that a process (e.g., B) is always tested after all of its descendants (in case of B: G, H, I). Therefore, if the anomaly appears when moving B to the end position, this can only be caused by B itself, but not due to descendants of B, since these have already been tested.

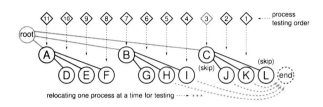

Fig. 5.4 LPs placed at process queue end (stage 4a)

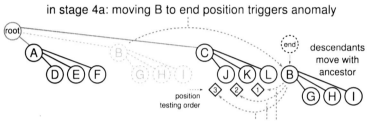

Fig. 5.5 LP and descendants ripple back (stage 4b)

Second dependent process (stage 4b). Once an LP with dependencies has been identified in Stage 4a, it is moved step-by-step starting from the end back towards its original position. It is advantageous to start from the end position, since this again ensures that the first dependent LP is always already tested against all descendants of possible second dependent LPs. In Fig. 5.5, an example is shown where B has been identified as first dependent process. During the following simulation runs, B will be moved step-by-step back towards its original position until the anomaly disappears, as indicated by the numbers inside the diamond marks. The LP moved over last must be the second dependent process. Since it is already known from stage 4a that the second dependent process must be located between the first dependent LP and the end position, it is sufficient to move B step-wise towards the start until one step before reaching its original position.

Once both processes have been identified, the testing framework can be instructed to run the simulation again, reversing the order of the two dependent processes. Just before executing the offending LP invocations, a hook function is called that can be used to place debugger breakpoints to allow manual inspection of the LP states.

5.5 Experiments and Case Studies

In this section, the practicability of *SCandal* is evaluated. First, the results of a number of synthetic tests are given in Sect. 5.5.1. After that, the experiences of testing SoClib [23] systems are summarized in Sect. 5.5.2. Finally, the results of

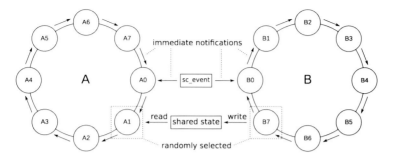

Fig. 5.6 Corner cases – multiple invocation test

analyzing a large system compatible with parallel SystemC simulation are presented in Sect. 5.5.3.

5.5.1 Synthetic Tests

Random automated synthetic tests. were generated consisting of randomly generated SDTs with randomly selected dependent LPs appearing in random delta-cycles. Dependent LPs simulate non-determinism by accessing a shared global variable. The first LP sets this variable from its initial value of 0 to 1. The second LP reads this value and saves it inside local storage for deferred output. *SCandal* has successfully completed all of more than 10^6 test iterations.

Corner cases. not covered by the random tests have been checked using hand-crafted tests. Such corner cases include PEO dependencies between LPs that run multiple times during a single delta-cycle and more particularly LPs exercising self-reactivation. Since the OSCI SystemC kernel does not allow an LP to be scheduled again while it is running, such behavior is only valid under special circumstances as described below.

For an LP to be able to re-run itself during a single delta-cycle, a second LP must be introduced that gets woken up from the first LP using an immediate notification. It is the task of the second LP to wake up the first LP also using an immediate notification before yielding its execution.

Figure 5.6 illustrates a test simulator has been created to investigate such kind behavior. It consists of two process groups with eight LPs each. In each group, the first LP is woken up at the beginning of the simulation and triggers the next one from its own group using an immediate notification. The last LP from each group notifies the first LP again. This circle continues for a random number of full iterations. From each group one LP is randomly selected to expose nondeterministic behavior by accessing a shared global variable. The test is passed once both LPs that are responsible for the anomaly have been identified and the number of invocations

Table 5.2 OSCI TLM 2.0 examples results

Test	$c_{observation}$	$c_{causation}$	Runtime (ms)	RefDB size (kB)	c_{end}
bus	1	0	287	106	600
cancel-all	76	76	119	26	101
multi-sockets	1	0	164	25	159
at-mixed-targets	2	1	111	393	868
at-ooo	3	1	101	438	931
at-4-phase	4	1	98	456	1,026
at-ext.-opt.	2	1	99	461	962

that these LPs went through before triggering the nondeterminism has been stated correctly. *SCandal* has successfully completed all of more than 10^5 test iterations.

All OSCI SystemC self-tests. have been analyzed, and at least in the original, unmodified tests no anomalies were found. The self-tests are coded either in a fully deterministic modeling style, or do not contain any concurrency at all (e.g., testing output formatting).

sc_event_queue. is a standard primitive channel which buffers multiple event notifications for a single event. The OSCI implementation (version 2.2.0) contains PEO dependencies. An assertion violation is triggered in case the queue *cleanup* process is scheduled *after* another process (LP_{cancel}) that calls the *cancel_all* method of the event queue. This situation can be created by introducing an event that activates LP_{cancel} and then notifying this event *before* notifying the event queue.

This issue appeared when trying to adapt the OSCI event queue for parallel simulation. While it is straight-forward to protect its data members with synchronization primitives, diagnosing the infrequently appearing assertion violations due to PEO variations in parallel simulation mode was a complex task. *SCandal* now reproducibly triggers the assertion during stage two in sequential mode.

OSCI TLM 2.0 examples. are written using a nondeterministic modeling style. In Table 5.2, observation cycles ($c_{observation}$), causation cycles ($c_{causation}$), the required wallclock time to run the complete analysis (Runtime), the size of the reference observation database (RefDB) required as well as the number of delta-cycles of the individual tests (c_{end}) are summarized for all tests with anomalies detected by the LP scheduling or standard output observer.

The main cause for nondeterministic behavior is that inside the interconnect models requests are forwarded without being arbitrated.

SCOOT examples [21]. *pressure* and *mutex* have been analyzed. Table 5.3 summarizes the analysis results.

In the *mutex* example, a dependency is detected between two LPs that concurrently access a shared variable protected by a mutex lock.

In the *pressure* example, two LPs perform concurrent operations on a shared variable. To identify the nondeterministic behavior, the proposed tool required a custom

Table 5.3 SCOOT examples results

Test	$C_{observation}$	$C_{causation}$	Runtime (ms)	RefDB size (kB)	c_{end}
mutex	0	0	8	82	400
pressure	–	–	<4	3.3	600
pressure-obs.	0	0	4	5.7	40

observer (see Sect. 5.4.1) for the shared variable to detect the nondeterminism, since its value is never further used or output.

Indexer [12]. is a well-known benchmark for MC tools. For analysis with the proposed tool, the already present observation instrumentation for POR tools had to be replaced with plain variable accesses. Further on, after the test finishes, the computed hash table needed to be output by a newly introduced process once so that the result of the computation is visible to the LP output observer. *SCandal* was then able to automatically identify the nondeterministic test behavior and the involved processes.

5.5.2 SoClib

SoClib is a model library for virtual MPSoC prototyping [23]. The SoClib example systems *tutorial0*, *tutorial0_ppc405* and *tutorial1* of the SoClib tutorial [24] containing MIPS and PowerPC processor models have been analyzed to evaluate how *SCandal* integrates with large projects: It was sufficient to exchange the SystemC kernel. Nondeterminism analysis yielded no anomalies though, even after manually instrumenting the simulations with custom observers. This was expected, as the tested SoClib models are written using a cycle-accurate and bit-accurate (CABA) modeling style.

5.5.3 Parallel Simulation of Mixed-Level Multicore Platform

The original motivation to create the proposed tool stems from experiences during the development of the inhouse virtual Pitaya multicore platform (see Fig. 5.7). Its main use is experimentation with new simulation technologies. It comprises cycle-accurate (CA) as well as TLM-style models and behaves deterministically. The CA inhouse processor models were generated using [25].

Two times nondeterminism anomalies were encountered when executing in parallel mode, which manifested themselves in minor variations in the timing of data packets. Both times the issues were related to performance optimizations inside a newly developed queue modeling primitive used for deterministic message delivery to LPs. Although the issues have already been fixed by now, they serve as test cases for the proposed tool.

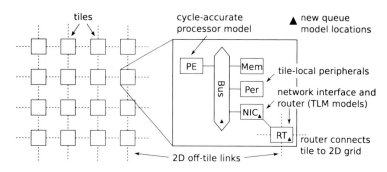

Fig. 5.7 Overview of Pitaya multicore platform

Table 5.4 Pitaya test results (8 × 9 tiles system)

Test	$C_{observation}$	$C_{causation}$	Runtime (min)	RefDB size (GB)	c_{end}
bug1	2,69,317	2,69,317	94	1.3	300 k
bug1-obs.	2,52,237	2,52,237	86	1.5	300 k
bug2	2,52,237	2,52,235	82	1.3	300 k
current	–	–	17	4.7	1 M

When testing the system using *SCandal* with both issues present, the default memory observer (see Sect. 5.4.1) detected a PEO dependency inside the new queue modeling primitive and correctly identified the involved LPs. Adding a custom observer to track activity details of the identified object allowed to detect the same anomaly with identical LPs even earlier in simulation time. For the second issue, the default observers already indicated the same processes and the same time as the custom observer.

In Table 5.4, the analysis results of a Pitaya system exhibiting both issues are shown. The system comprises 72 complete computational tiles. The resource consumption of the tool is still low enough to successfully complete all four stages of the analysis.

In retrospective, placing mutex locks to achieve thread-safety was a straightforward task for the given system. However, detecting and analyzing the PEO dependencies manually at that time was a complex and time-consuming activity. In contrast, today *SCandal* identifies the involved LPs reproducibly and automatically.

5.6 Conclusion and Outlook

The presented framework detects process execution order dependencies in a large variety of SystemC simulations. While following a testing methodology, the most important class of nondeterminism anomalies is reliably identified. *SCandal* is not intended to substitute MC or POR approaches. But, by reason of its low resource consumption, the tool can be employed instead in situations where it is not viable to use MC or POR tools due to hardware or computation time requirements.

In the future, parallel execution of reference and detection run will be enabled to compare observation data on the fly and to allow analysis runs of arbitrary length. Moreover, possible algorithms with logarithmic complexity for stages 4a and 4b need to be investigated to further reduce the time needed for analysis by *SCandal*.

Acknowledgements This work has been supported by the German excellence cluster UMIC and the European FP7 project EURETILE.

References

1. Blanc, N., Kroening, D.: Race analysis for SystemC using model checking. In: Proceedings of the 2008 IEEE/ACM International Conference on Computer-Aided Design, ICCAD'08, San Jose, pp. 356–363. IEEE Press, Piscataway (2008)
2. Chen, W., Domer, R.: An optimizing compiler for out-of-order parallel ESL simulation exploiting instance isolation. In: Proceedings of the 17th Asia and South Pacific Design Automation Conference (ASP-DAC), Sydney, pp. 461–466 (2012)
3. Cimatti, A., Micheli, A., Narasamdya, I., Roveri, M.: Verifying SystemC: a software model checking approach. In: Proceedings of the 2010 Conference on Formal Methods in Computer-Aided Design, FMCAD'10, Austin, FMCAD Inc. (2010)
4. Cockx, J., Denolf, K., Vanhoof, B., Stahl, R.: SPRINT: a tool to generate concurrent transaction-level models from sequential code. EURASIP J. Appl. Signal Process. **2007**(1), 213–213 (2007). http://dl.acm.org/citation.cfm?id=1289174
5. Dömer, R., Chen, W., Han, X., Gerstlauer, A.: Multi-core parallel simulation of system-level description languages. In: Proceedings of the 16th Asia and South Pacific Design Automation Conference, ASPDAC'11, Yokohama. IEEE Press, Piscataway (2011)
6. Godefroid, P.: Software model checking: the VeriSoft approach. Form. Methods Syst. Des. **26**, 77–101 (2005)
7. Große, D., Kühne, U., Drechsler, R.: HW/SW co-verification of embedded systems using bounded model checking. In: Proceedings of the 16th ACM Great Lakes Symposium on VLSI, GLSVLSI'06, Philadelphia, pp. 43–48. ACM, New York (2006)
8. Grotker, T.: System Design with SystemC. Kluwer, Norwell (2002)
9. Helmstetter, C., Maraninchi, F., Contoz, L.M., Moy, M.: Automatic generation of schedulings for improving the test coverage of systems-on-a-chip. In: Proceedings of the Formal Methods in Computer Aided Design, FMCAD'06, San Jose, pp. 171–178. IEEE Computer Society, Washington, DC (2006)
10. Herrera, F., Villar, E.: Extension of the SystemC kernel for simulation coverage. In: Forum on Specification and Design Languages, FDL'06, pp. 161–168 (2006)
11. IEEE Standard SystemC Language Reference Manual. IEEE Std 1666–2005 pp.0–1423 (2006). doi: 10.1109/IEEESTD.2006.99475 http://ieeexplore.ieee.org/stamp/stamp.jsp?tp=&arnumber=1617814&isnumber=33906
12. Indexer Benchmark: http://trac.assembla.com/scrv/browser/examples/rvs/indexer/indexer.cpp. Accessed 2011
13. Lu, S., Park, S., Seo, E., Zhou, Y.: Learning from mistakes: a comprehensive study on real world concurrency bug characteristics. In: Proceedings of the 13th International Conference on Architectural Support for Programming Languages and Operating Systems, ASPLOS XIII, Seattle, pp. 329–339. ACM, New York (2008)

14. Magnusson, P.S., Christensson, M., Eskilson, J., Forsgren, D., Hallberg, G., Hogberg, J., Larsson, F., Moestedt, A., Werner, B.: Simics: a full system simulation platform. Computer **35**(2), 50–58 (2002)
15. Mello, A., Maia, I., Greiner, A., Pecheux, F.: Parallel simulation of SystemC TLM 2.0 compliant MPSoC on SMP workstations. In: Proceedings of the Conference on Design, Automation and Test in Europe, DATE'10, pp. 606–609, Leuven. European Design and Automation Association (2010)
16. Musuvathi, M., Qadeer, S., Ball, T., Basler, G., Nainar, P.A., Neamtiu, I.: Finding and reproducing Heisenbugs in concurrent programs. In: Proceedings of the 8th USENIX Conference on Operating Systems Design and Implementation, OSDI'08, Berkeley, pp. 267–280. USENIX Association (2008)
17. Nanjundappa, M., Patel, H.D., Jose, B.A., Shukla, S.K.: SCGPSim: a fast SystemC simulator on GPUs. In: Proceedings of the 2010 Asia and South Pacific Design Automation Conference, ASPDAC'10, Taipei. IEEE Press, Piscataway (2010)
18. Open SystemC Initiative: OSCI TLM-2.0 Language Reference Manual. http://www.accellera. org/downloads/standards/systemc (July 2009)
19. Robert E. Lantz: Parallel: Scalability and performance for large system simulation. Ph.D. thesis, Computer Systems Laboratory, Stanford University (June 2007)
20. Schumacher, C., Leupers, R., Petras, D., Hoffmann, A.: parSC: synchronous parallel SystemC simulation on multi-core host architectures. In: Proceedings of the 8th IEEE/ACM/IFIP International Conference on Hardware/Software Codesign and System Synthesis, CODES/ISSS'10, Scottsdale, pp. 241–246. ACM, New York (2010)
21. SCOOT, a tool for the static analysis of SystemC http://www.cprover.org/scoot. Accessed 2011
22. Sen, A., Ogale, V., Abadir, M.S.: Predictive runtime verification of multi-processor SoCs in SystemC. In: Proceedings of the 45th Annual Design Automation Conference, DAC'08, Anaheim, pp. 948–953. ACM, New York (2008)
23. SoClib, an open platform for virtual prototyping of multi-processor systems on chip. http:// www.soclib.fr. Accessed 2011
24. SoClib Appliance. http://www.soclib.fr/appliance/soclib-vm-latest.zip
25. Synopsys Inc.: Synopsys processor designer. http://www.synopsys.com/Systems/BlockDesign/ ProcessorDev/Pages/default.aspx
26. The GNU Linker ld.: http://sourceware.org/binutils/docs/ld/Options.html. Accessed 2011

Chapter 6
A Design and Verification Methodology for Mixed-Signal Systems Using SystemC-AMS

Yao Li, Ramy Iskander, Farakh Javid, and Marie-Minerve Louërat

Abstract This chapter presents a unified platform for design and verification of mixed-signal systems using SystemC-AMS standard. The platform bases on a bottom-up design and a top-down verification methodologies. In the methodologies, several hierarchical abstraction levels of the system are considered. These abstraction levels are: *system, functional, macromodel* and *circuit* levels. We introduce a simple and efficient solution to implement an interface between system level models and their circuit level realizations. Simulation tools such as SystemC-AMS and Spice simulators are combined with a sizing tool named CHAMS, in order to achieve a unified and standard design and verification environment. Moreover, a transient simulation scheme is proposed to simulate nonlinear dynamic behavior of complete mixed-signal systems. The unified platform is used to design and verify a pipeline ADC. The simulation results prove the effectiveness of the proposed structure and methodology.

6.1 Introduction

Today's electronic systems became more and more complex and heterogenous. Besides digital parts performing signal processing operations and software task running on dedicated processors, analog and mixed-signal parts became very critical components in most electrical systems. Due to their complexity and design challenges [1–3], these components became a bottleneck in the design process of a SoC. On the one hand, important system functions like clock generation, or signal conversion between analog and digital signals, are realized by analog circuits. On the other hand, analog circuits are difficult to design and reluctant to design

---

Y. Li (✉) • R. Iskander • F. Javid • M.-M. Louërat
Université Pierre et Marie Curie, LIP6, 4 Place Jussieu, 75252 Paris, France,
e-mail: Yao.Li,Ramy.Iskander; Farakh.Javid; Marie-Minerve.Louerat@lip6.fr

J. Haase (ed.), *Models, Methods, and Tools for Complex Chip Design*, Lecture Notes in Electrical Engineering 265, DOI 10.1007/978-3-319-01418-0__6,
© Springer International Publishing Switzerland 2014

automation. Right now, various tools are employed to simulate these components from system-level specifications down to circuit-level realizations. For example, SystemC-AMS [4] uses analog and mixed-signal extensions to define a system-level behavior, while SPICE [5] simulators are used to simulate transistor netlists to determine linear and nonlinear circuit behavior.

In [6], a method was proposed to achieve a top-down analog design methodology. Nevertheless, this attempt to establish a link between system-level description and circuit-level realization combines many different design tools to achieve a common design environment. This lacks a clear implementation of a unified platform supporting an interface to connect system-level descriptions to its circuit-level realization. Today analog circuit design demands further automation, hence the co-simulation of mixed-signal design is overly used in recent research. It aims at mixing with behavioral models, such as transfer function, VHDL or verilog behavioral descriptions and signal flow graphs, with structural model, such as circuit netlist (resistor, inductance, capacitor, transistor, source elements).

Many existing co-simulation approaches are based on SystemC [7] or SystemC-AMS [4]. In [8], co-simulation refined models with Timed Data Flow (TDF) paradigm of SystemC-AMS are presented. SystemC-AMS acts as master controlling VHDL test bench. Another attempt to achieve analog mixed signal simulation using loose coupling between SystemC and SPICE is presented in [9]. The SPICE simulator is restarted till next SystemC event using restored simulation state from previous call.

In this chapter, we propose a unified platform for design and verification of mixed signal systems which relies on the bottom-up design process. This bottom-up process of synthesis and optimization is followed by a top-down analysis and simulation. We provide a standard interface between system-level models and their circuit-level realizations to be able to connect and simulate different components with different levels of abstraction. Our approach encapsulates the full spectrum of design from the system-level to the circuit-level of a mixed-signal design in one platform based on C/C++ and SystemC. Having a single tool for design and simulation is more than convenient, since it helps in optimizing efficiency in the design chain.

This work is part of CHAMS project [10–13], which is developed at the LIP6 (Laboratoire d'informatique de Paris 6). CHAMS is an analog design automation tool that provides assistance to the designer for the sizing and biasing step, as well as for the layout generation step. Three advantages that make CHAMS suitable to be an efficient system to circuit interface:

1. It is based on the C/C++ language which can be used with SystemC/SystemC-AMS.
2. CHAMS implements a standard interface for the encapsulation of an electrical simulator.
3. It has a very fast sizing and biasing tool [10–12].

6 A Design and Verification Methodology for Mixed-Signal Systems

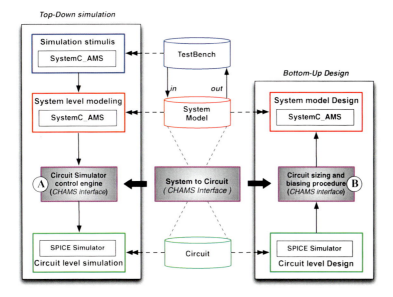

Fig. 6.1 Proposed platform architecture

The chapter is organized as follows. Section 6.2 describes the AMS extensions of SystemC, the circuit design procedure and the circuit simulator control engine that are part of the unified platform. Section 6.3 gives an overview of the different levels of abstractions by introducing a pipeline ADC case study. The system level to circuit level interface implementation is introduced in Sect. 6.4. Section 6.5 presents the nonlinear dynamic analog simulation method. The simulation results are reported in Sect. 6.6. Section 6.7 concludes the demonstrated work.

6.2 Unified Platform Architecture

Figure 6.1 represents the proposed platform architecture to connect system-level descriptions to their circuit-level realizations. This platform is composed of a *bottom-up design* path as well as a *top-down simulation* path.

1. The bottom-up design path consists of :
   (a) A SPICE simulator used for sizing.
   (b) The SPICE simulator is controlled by the circuit sizing and biasing procedure.
   (c) The sizing and biasing procedure is called from the SystemC-AMS models. This procedure provides a sized circuit to be used in the following top-down simulation.

2. The top-down simulation path consists of :

(a) The *testbench* instantiates the SystemC-AMS models and generates the corresponding stimuli signals.
(b) SystemC-AMS models to be simulated.
(c) Circuit simulator control engine is controlled by CHAMS and pass the stimulus to the circuit netlist.
(d) Analysis simulator running the circuit netlist.

As shown in Fig. 6.1, CHAMS interface consists of two main parts: *the circuit simulator control engine* (block A in Fig. 6.1) and *the circuit sizing and biasing procedure* (block B in Fig. 6.1).

A complete system can be described using only the AMS extension of SystemC [4] with some parts described in SPICE netlists. The proposed platform is capable to simulate the whole system with different levels of abstraction. With this method, we can verify the impact of a circuit block (transistor netlist) at the system level. Along with it, the system level considerations traverse the different design levels from the system level over the functional level down to the circuit level.

6.2.1 SystemC AMS Extensions

The SystemC AMS extensions [4, 14] provide a framework for functional modeling [15], integration validation, and virtual prototyping [16] of *Embedded Analog/Mixed-Signal Systems*. The SystemC AMS extensions provide three different models of computation: Timed Data Flow (TDF), Linear Signal Flow (LSF), and Electrical Linear Networks (ELN).

In our approach, the Timed Data Flow model of computation is used. Unlike the TDF modeling style, the LSF and ELN modeling styles can only be composed from their own linear primitives, which are too limited to model complex mixed-signal circuit. TDF is a discrete-time modeling style, which considers data as signals sampled in time. These signals are tagged at discrete points in time and carry discrete or continuous values, such as voltage amplitudes. Besides, TDF can be used with great efficiency to model complex non-conservative behaviors at system, functional and macromodel level. Figure 6.2 shows the principle of the TDF modeling. The basic entities found in the TDF model of computation are: the TDF modules, the TDF ports and the TDF signals. The set of connected TDF modules form a directed graph, called a TDF cluster as defined below :

- TDF modules are the vertices of the graph.
- TDF signals correspond to its edges.

Each TDF module involved in the cluster contains a specific C++ member function, named **processing**(), that computes a value at each time step.

Fig. 6.2 A basic TDF model with 3 TDF modules and 2 TDF signals

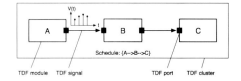

The TDF modules' schedule and sampling rate are known in advance for each TDF cluster. Therefore, this schedule can be statically determined before the simulation starts and corresponds to a static schedule of the TDF cluster.

If enough data samples are available at its input ports, depending on the involved port rates, the samples computed by a TDF module are written to the related output ports and describe continuous-time behaviors.

6.2.2 CHAMS Sizing and Biasing Engine

CHAMS [10–12] is a tool that provides assistance to the designer for the design of analog firm intellectual properties (IP) [17, 18]. It allows to generate the analog IP sizing and biasing procedure. It consists of the following three parts: *sizing and biasing operators, graph representation, simulator encapsulation*.

6.2.2.1 Sizing & Biasing Operators

A reference transistor is defined for each device as the only transistor to be sized and biased in the device. Sizing and biasing operators aim at computing the sizes and biases of reference transistors, they are based on the numerical inversion of the transistor compact model. Every transistor is defined by the following parameters: W (width), L (length), V_{GS} (gate-source voltage), V_{DS} (drain-source voltage), V_{BS} (bulk-source voltage), V_{EG} (overdrive gate voltage), I_D (drain current), $Temp$ (temperature). Each operator has a set of input parameters that are set by the designer and computes unknown widths and biases (see Table 6.1, where $V_{EG} = V_{GS} - V_{TH}$). An operator computes either:

$$W = f_W(Temp, I_D, L, V_{GS}, V_{DS}, V_{BS}) \tag{6.1}$$

or:

$$V_{GS} = f_{V_{GS}}(Temp, I_D, W, L, V_{DS}, V_{BS}) \tag{6.2}$$

Table 6.1 Class definition of sizing & biasing operators

Operator	Definition
$OPVS(V_{EG}, V_B)$	$(Temp, I_{DS}, L, V_{EG}, V_D, V_G, V_B) \mapsto (V_S, W, V_{TH})$
...	...
$OPVG(V_{EG})$	$(Temp, I_{DS}, L, V_{EG}, V_D, V_S) \mapsto (V_G, W, V_B, V_{TH})$
...	...
$OPVGD(V_{EG})$	$(Temp, I_{DS}, L, V_{EG}, V_S) \mapsto (V_G, V_D, W, V_B, V_{TH})$
...	...
$OPW(V_G, V_S)$	$(Temp, I_{DS}, L, V_D, V_G, V_S) \mapsto (W, V_B, V_{TH})$
...	...
$OPIDS(V_G, V_S)$	$(Temp, W, L, V_D, V_G, V_S) \mapsto (I_{DS}, V_B, V_{TH})$
...	...

f_W and $f_{V_{GS}}$ are two inverse functions of the transistor compact model given in Eq. (6.3):

$$I_D = f_{MODEL}(Temp, W, L, V_{GS}, V_{DS}, V_{BS}) \tag{6.3}$$

where *MODEL* is a standard transistor model like BSIM3v3 [19], BSIM4 [19], PSP [20] and EKV [21]. f_W and $f_{V_{GS}}$ are monotonic functions, thus Eqs. (6.1) and (6.2) are solved using the Newton-Raphson method.

Table 6.1 gives the definition of the main five classes of the sizing and biasing operators applied to the MOS transistor. Let us examine in further detail one operator such as $OPVG(V_{EG})$. The *OPVG* operator class is *gate voltage* V_G. The $OPVG(V_{EG})$ is called whenever V_{EG} is known and the MOS transistor is bulk-source connected. This operator computes V_G, V_B, V_{TH} and W, simultaneously, in terms of $Temp$, I_{DS}, L, V_{EG}, V_D and V_S.

6.2.2.2 Graph Representation

To size and bias a reference transistor, a bipartite DAG (Directed Acyclic Graph) is associated with it. The bipartite graph [22] for the sizing and biasing of the diode-connected transistor using operator $OPVGD(V_{EG})$ (Table 6.1) is shown in Fig. 6.3b. A set of input parameters are defined for the diode-connected transistor. The sizing and biasing operator $OPVGD(V_{EG})$ is then called to compute the set of output parameters.

6.2.2.3 Simulator Encapsulation

Sizing and biasing operators use a specific simulator encapsulation that allows to interface with industrial design kits to ensure very accurate computed results. The simulator encapsulation is illustrated in Fig. 6.4. At the bottom is an electrical netlist that specifies the suitable technology and contains only two transistors: one

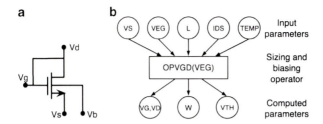

Fig. 6.3 (a) NMOS reference transistor. (b) Graph representing the input parameters and output parameters of the operator OPVGD

Fig. 6.4 CHAMS sizing engine: electrical simulator encapsulation within sizing and biasing operators

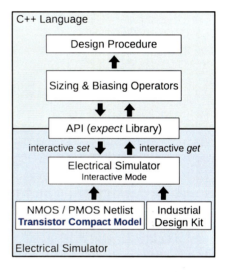

PMOS and one NMOS, entirely sizable and biasable through simulator interactive commands. It is loaded by the electrical simulator launched in interactive mode. Three types of interactive commands are evaluated: *set*, *get* and *run*. The first one allows to set all transistor known parameters (sizes and biases) inside the simulator. The second one enables to get all currents, voltages and small signal parameters computed by the simulator. After a set command, a simulation must be run using run command, in order to compute the DC operating point of the transistor. An API is developed using *expect* library [23] to automate *set*, *get* and *run* commands execution using simulator interactive mode. Sizing and biasing operators are optimized to minimize the number of calls to the simulator, which can reach several thousands during sizing.

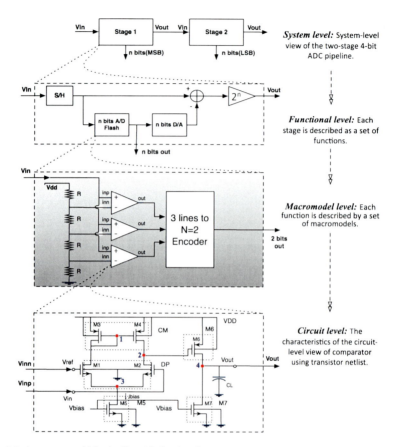

Fig. 6.5 A two stage 4-bit pipeline ADC using four levels of abstractions

6.3 Proposed Levels of Abstraction

In this section, we present the different levels of abstraction used in a two stage 4-bit pipeline analog-to-digital converter (ADC) design hierarchy. The ADC is shown in Fig. 6.5, where the hierarchical design view is organized into four levels of abstraction: *system, functional, macromodel* and *circuit*.

- The *system level* represents the highest abstraction level. The system is modeled as a monolithic unit able to process the specified functionalities. In our case, it describes the system view of the two-stage 4-bit ADC pipeline. It is composed of two stages, where the MSBs come from the first stage, and the LSBs from the second one.
- The *functional level* describes each stage as a set of mathematical functions, algorithms or state presentation. The block diagram shows that the analog input V_{in} is first sampled and held, quantized to ($n = 2$) bits by the A/D Flash and fed

to a ($n = 2$) bits D/A, whose output is subtracted from the input. The resulting residue is amplified by a factor 4 before stepping into the next stage.
- The *macromodel level* is the hierarchical level where some function are described by a set of macromodels. For example, the 2-bit resolution of the Flash ADC function are obtained by using a resistor ladder with 2^n resistors and $2^n - 1$ comparators. Macromodels [24] use simplified simulation elements and mathematical functions to define a specific behavior of a given function. The output of each comparator is a high/low value signal, then a logic encoder is used to convert the signal to a binary coded word.
- The *circuit level* describes the characteristics of the circuit using SPICE netlist. It is implemented with the typical basic analog circuits like current sources, differential pairs, current mirrors, output buffers, ... [25]. The comparator macromodel is implemented as a two-stage structure [26] as shown in Fig. 6.5.

6.4 Implementation of the Unified Platform

The unified design and verification platform is presented in Fig. 6.6. It is composed of three parts: the SystemC-AMS environment, the CHAMS interface and the two electrical simulators (sizing simulator and analysis simulator). In the SystemC-AMS simulator, a set of TDF modules are organized to build the two stage 4-bit pipeline ADC with the four hierarchical levels shown in Fig. 6.5. Each TDF module is integrated in a separate file. For instance, a module named *source_constant.h* works as a resistor ladder circuit for generating the reference voltage.

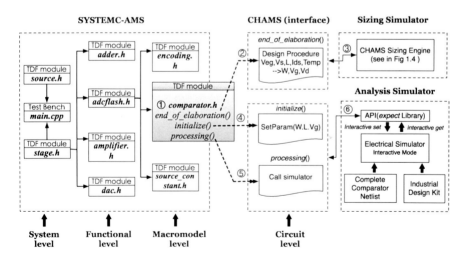

Fig. 6.6 The unified design and verification platform with SystemC-AMS. The four hierarchical levels shown in Fig. 6.5 are implemented in the design platform

Fig. 6.7 Algorithm that permits to realize the CHAMS interface from system-level to circuit-level

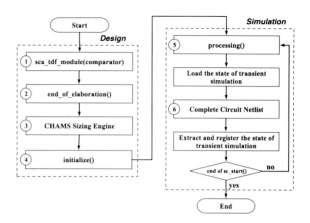

As shown in Fig. 6.6, the CHAMS interface is related to a TDF module named *comparator.h*. The analysis simulator (step 6 in Fig. 6.6) is encapsulated in CHAMS to load the whole comparator circuit netlist. The test bench is described in the file *main.cpp*, using a TDF module *source.h*, which generates the input signal.

The algorithm that allows to implement the CHAMS interface (blocks A and B in Fig. 6.1) from system-level to circuit-level is represented in Fig. 6.7. In this algorithm, each step is defined by a number that corresponds to either a TDF module (step 1) or a function call (steps 2–6) shown in Fig. 6.6.

This algorithm can be separated to two parts, which are *system design* and *system simulation* respectively. The first four steps in the algorithm are executed only once, they correspond to the sizing and biasing of the comparator within the complete system-level description of the two-stage pipeline ADC, it is the design part in Fig. 6.1. In step 2, the sizing and biasing procedure is executed by using a electrical simulator named *SizingSimulator* (see line 18 in Sect. 6.4.2). This sizing simulator is closed at the end of step 3.

From step 5 until the end of execution, they correspond to the system simulation including the circuit level comparator. It is the simulation and verification part in Fig. 6.1. In steps 5 and 6, at each *timestep*, the signal interface passes the input samples and evaluate the simulated output samples. These steps are executed until the last input sample is processed. At the first execution of step 5, a single electrical simulator named *AnalysisSimulator* (see line 21 in Sect. 6.4.2) is opened, it calls the complete circuit netlist at the step 6, this analysis simulator is closed at the end of the system simulation. During the simulation, a loading and saving of the state of the transient simulation for each time step is performed before and after step 6, these two blocks refer to the initial condition of the circuit simulation (see lines 37, 39 in Sect. 6.4.4). The following sections will detail the unified design platform.

6.4.1 Comparator TDF Module

The comparator TDF module (implemented in *comparator.h*, lines 7–16). The comparator is implemented as TDF module. The implementation of the constructor of the comparator is described below:

- Lines 1, 2: Define the parameters used to size and bias the comparator.
- Lines 3, 4: Define the parameters used to store simulations results.
- Lines 5, 6: Define the parameters used to store initial conditions for transient behavior at each time step.
- Lines 8, 9: Define two TDF input ports and one TDF output port that carries the continuous-value (real) signals [4, clause 2.2.2].
- Line 10: The constructor expects the parameters by using the const reference to an object.
- Lines 11–13: Define the pointer of allocated objects for each struct, which are initialized by the constructor.

```
1    struct sizing_parameters{
2          double Temp, Veg, L, Vs, Ids ,...; };
3    struct analysis_parameters{
4          double W, L, Vg ,...; };
5    struct initial_conditions{
6          double Init_1, Init_2, Init_3, Init_4; };
7    SCA_TDF_MODULE(comparator) {
8          sca_tdf::sca_in<double> inp, inn;
9          sca_tdf::sca_out<double> out;
10         comparator(
                   const sizing_parameters& ,
                   const analysis_parameters& ,
                   const initial_conditions& );
11         sizing_parameters *sizing_param;
12         analysis_parameters *analysis_param;
13         initial_conditions *initial;
14         void end_of_elaboration();
15         void initialize();
16         void processing();
       };
```

6.4.2 end_of_elaboration() function

This callback function is called at the very end of elaboration before starting simulation [7, clause 4.4.2]. This callback will be used to instantiate two simulator objects. The first simulator object will be used to size and bias the comparator before the start of simulation. The second simulator object will be used to simulate the comparator netlist within the complete two-stage ADC system simulation.

The implementation of the design procedure (using sizing and biasing operators) is described below:

- Line 18: Defines an electrical simulator object for the circuit sizing and biasing procedure (see Fig. 6.4).
- Line 19: The sizing and biasing simulator loads the netlist which is related to transistor compact model.
- Line 20: Defines an electrical simulator object for the simulation of the complete circuit netlist (comparator).
- Line 21: The analysis simulator loads the complete netlist which contains the circuit information of the comparator.
- Line 22: Defines a circuit object into the sizing and biasing procedure.
- Lines 23, 24: Define the operators to compute the sizes and biases of reference transistors with different operator classes (see Sect. 6.2.2.1).
- Lines 25, 26: Add the devices to the circuit.
- Line 27: Sets the input parameters for the sizing operators. The input parameters are loaded from a pointer object of struct named *sizing_param* (line 11 in Sect. 6.4.1).
- Line 28: Runs the design procedure with the pre-defined sizing simulator and operators.
- Line 29: Gets the output parameters computed by the sizing procedure. The output parameters are stored in a pointer object of struct named *analysis_param* (line 12 in Sect. 6.4.1).

```
17   void comparator::end_of_elaboration() {
18       ElectricalSimulator *SizingSimulator
             = new ElectricalSimulator;
19       SizingSimulator->loadnetlist ("nmos-pmos.spi");
20       ElectricalSimulator *AnalysisSimulator
             = new ElectricalSimulator;
21       AnalysisSimulator->loadnetlist ("comparator.spi");
22       Circuit *CirComparator = new Circuit;
23       Operator *OperatorCM= new Operator (OPVGD);
24       Operator *OperatorDP= new Operator (OPVS);
         ... ...
25       CirComparator->addDevice(OperatorCM);
26       CirComparator->addDevice(OperatorDP);
```

```
         ...  ...
27       SetInputParameters(sizing_param);
28       SizingProceduce(SizingSimulator, CirComparator);
29       GetOutputParameters(analysis_param);
   }
```

6.4.3 initialize() function

The member function *initialize()* is called once-only after the callback of the member function **start_of_simulation** and before the first call to the member function **processing** of a TDF module. In our proposed methodology, this member function is used to set the variables for the circuit simulation [4]. Following the computation and registration of the analysis parameters in lines 28–29, the *SetSimuParam* function (lines 31–33) will automatically transmit the parameter values to the simulator used to simulate the complete circuit netlist (*AnalysisSimulator*).

```
30  void comparator::initialize() {
31      AnalysisSimulator->SetSimuParam
            ("Width_M1",analysis_param->W);
32      AnalysisSimulator->SetSimuParam
            ("Length_M1",analysis_param->L);
33      AnalysisSimulator->SetSimuParam
            ("Vg_M1",analysis_param->Vg);
        ...  ...
   }
```

6.4.4 processing() function

The member function **processing()** is the only mandatory function that needs to be overloaded in any TDF module, its is described below:

- Lines 35, 36: Samples are read from a TDF module input port by calling its member function **read**, then the sample values are passed to the complete circuit netlist with the name V_{in}, V_{ref}.
- Line 37: Sets the initial condition for transient simulation, this simulation method is described in Sect. 6.5.
- Line 38: Performs comparator transient simulation.
- Line 39: Gets the initial condition from transient simulation, this simulation method is described in Sect. 6.5.

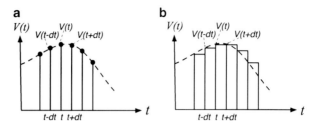

Fig. 6.8 (**a**) TDF signal with sampled values. (**b**) Transient simulation with a set of pulse signals

- Line 40: The output samples are extracted from the electrical simulator and are written to the TDF module output port using **write** function of TDF module.

```
34   void comparator::processing() {
35       AnalysisSimulator->SetSimuParam
             ("Vin" , inp.read());
36       AnalysisSimulator->SetSimuParam
             ("Vref" , inn.read());
37       AnalysisSimulator->SetInitialConditions
             (initial_conditions);
38       AnalysisSimulator->TransientAnalysis();
39       AnalysisSimulator->GetInitialConditions
             (initial_conditions);
40       out.write( GetExtractValue("Vout") );
     }
```

6.5 Transient Analysis Method

For the design of analog circuits, the most important characteristic is the consideration of signals that are continuous in time and value. We aim at performing conservative nonlinear simulations for the components described in SPICE netlist. Contrary to this, the Timed Data Flow (TDF) model of computation is not conservative, it considers values that are discrete in time and value. To be able to handle such problem, we convert the TDF input signal V_{inp} of the comparator shown in Fig. 6.8a to the sample and hold version shown in Fig. 6.8b. This conversion will be considered as the stimuli signal of the comparator during SPICE simulation. The following steps are performed during comparator simulation:

1. As shown in Fig. 6.8b, in the comparator netlist, a pulse signal with a voltage value of $v(t)$ is generated at time t. The pulse width is set to the sampling period dt (line 4 in the code below). At the beginning of the transient analysis, the

voltages at nodes 1, 2, 3, 4 marked in Fig. 6.5 are respectively set to ic1, ic2, ic3, ic4 (line 6). At the end of the transient simulation, the voltage at the four nodes are extracted to be the initial conditions for the next simulation (line 7). The four nodes connect to all the small-signal capacitances in the circuit.

```
1     .PARAM ic1 = ...
2     .PARAM dt = ...
3     .PARAM vin = ...
4     Vpulse 0  vin  0  dt
5     .TRAN  0.1n  dt  uic
6     .IC v(1)=ic1 v(2)=ic2 v(3)=ic(3) v(4)=ic4
7     .EXTRACT label = ic1 yval(v(1), dt)
      ... ...
```

2. In **processing**() function, the *SetInitialConditions* function (line 37 in Sect. 6.4.4) is called to set the initial conditions at the beginning of each simulation step. At the end of current simulation, the value of each node is retrieved using *GetInitialConditions* (line 39 in Sect. 6.4.4) to be used as the initial conditions for the simulation of the next time step. This function uses the simulator control engine to extract the initial conditions from the simulator using the predefined labels such as *ic1* in line 7 using the .EXTRACT command.

Using the above approach, the unified platform for mixed signal system design can mix non-conservative system-level behavior with conservative nonlinear circuit simulation.

6.6 Experimental Results

In this section, we use the unified platform to design and verify the two-stage 4-bit pipeline ADC shown in Fig. 6.5. The first step consists in implementing the TDF modules that constitute the system level view of the pipeline ADC. The second step consists in developing the sizing and biasing procedure for circuit level view of the comparator. The sizing procedure and simulation results are shown in the following parts.

All experiments have been carried out on a 32-bit Linux computer with 2 Duo Processor (3M Cache, 2.93 GHz), and 2 GB of memory.

6.6.1 Sizing and Biasing Procedure of the Two-Stage Comparator

The topology of two-stage comparator is shown in Fig. 6.5. The first stage consists of two basic devices: a current mirror, a differential pair. The purpose of the second

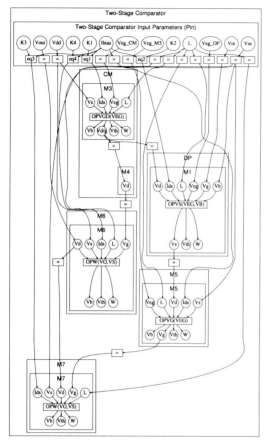

Fig. 6.9 Design view of the comparator circuit: (**a**) Bipartite graph (i.e. design procedure). P_{in} parameters are propagated to each operator input parameters. (**b**) Input parameters P_{in} (TEMP nodes are omitted) and output parameters P_{out}

stage is to get a much faster linear response. C_L is the input capacitance of the following block (encoder).

The sizing and biasing procedure of the two-stage comparator is shown in Fig. 6.9a with a 130 nm process. It is a *bipartite graph* that contains the designer knowledge to size and bias the comparator. The designer's knowledge is represented by P_{in} set of input parameters (at the top of the graph). Parameters in P_{in} (see in Fig. 6.9b) are propagated to each operator input parameters through equations and equalities. An example of equation is given with $eq3 : I_{ds\_M7} = K_3 \cdot I_{bias}$ where $\{K_3, I_{bias}\} \in P_{in}$. The resulting output parameters P_{out} are listed in Fig. 6.9b. The bipartite graph is a sequence of sizing and biasing operators, it is evaluated from top to bottom.

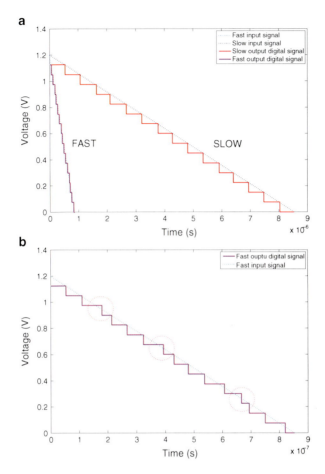

Fig. 6.10 Results of ADC simulations with sample rate (**a**) 30 and 300 MHz. (**b**) 300 MHz

6.6.2 Simulation Results of a Two-Stage Pipeline ADC

The testbench of the unified platform (see in Fig. 6.1) receives the input signals with the form $V(t) = V_{DD} - k \cdot t$ as a function of time t, it is generated by a TDF module with the following features:

$$\begin{aligned} N &= number\ of\ samples = 256 \\ T &= Simulation\ time = \frac{N}{sample\ rate} \\ k &= \frac{V_{DD}}{T} \end{aligned} \quad (6.4)$$

Two different input stimuli are applied to our design platform. The simulation results are shown in Fig. 6.10. The resolution of the pipeline ADC is 4 bits, since the number of quantization levels is equal to $2^4 = 16$, so $N/16$ simulated samples are used for each quantization. We fix the number of samples N. Then we change the simulation time using two different sample rates (30 and 300 MHz respectively) to obtain the same number of samples for each quantization level. In Fig. 6.10, we notice that by increasing the slope of the input signal, the precision of the pipeline ADC is decreased. Note that the higher slope corresponds to a faster varying signal. This observation exactly describes the relationship between relative speed and precision in a real pipeline ADC. The unified platform allowed to verify the performance of the pipeline ADC within the context of system-level and circuit-level description.

The total number of CPU-seconds that the process spent is less than 1 min, where the sizing and biasing procedure took less than 1 s.

6.7 Conclusion

This chapter presented a unified platform for design and verification of mixed-signal systems. It is based on a bottom-up design process and a top-down verification method, which efficiently provides a standard interface between system-level models and their circuit-level realizations. This interface is able to connect and simulate different components with different levels of abstraction.

The proposed framework interfaces SystemC-AMS and SPICE with CHAMS fast sizing and biasing. The framework is used to effectively design complex systems, and links to lower level circuit design techniques. A transient simulation scheme has been proposed to allow the simulation of system-level non-conservative models along with conservative circuit-level netlists. The platform has been used to design and simulate a two-stage pipeline ADC. The simulation time as well as simulation results prove the effectiveness of the proposed method.

Acknowledgements This work was funded by the project Verification For Heterogenous Reliable Design and Integration (VERDI), which is supported by the European Commission within the 7th Framework Programme for Research and Technological Development (FP7/ICT 287562).

References

1. Rutenbar, R.A.: Design automation for analog: the next generation of tool challenges. In: ACM International Conference on Computer-Aided Design, San Jose, pp. 458–460 (2006)
2. Taranovich, S.: Analog design in the 21st century: challenges, tools, and IC advances. http://www.edn.com (2012)
3. Graeb, H.E.: Analog Design Centering and Sizing. Springer, Heidelberg (2007)
4. Accellera Systems Initiative: SystemC AMS 2.0 standard. http://www.accellera.org/downloads/standards/systemc/ams/ (2013)

5. Vladimirescu, A.: The SPICE Book. Wiley, New York (1994)
6. Sommer, R., Rugen-Herzig, I., Hennig, E., Gatti, U., Malcovati, P., Maloberti, F., Einwich, K., Clauss, C., Schwarz, P., Noessing, G.: From system specification to layout: seamless top-down design methods for analog and mixed-signal applications. In: Design, Automation and Test in Europe, Paris, pp. 884–891 (2002)
7. IEEE Computer Society: 1666-2011 IEEE Standard SystemC Language Reference Manual. IEEE, 1666–2011.
8. Zaidi, Y., Grimm, C., Hasse, J.: On mixed abstraction, languages, and simulation approach to refinement with systemC AMS. EURASIP J. Embed. Syst. (2010). doi:10.1155/2010/489365
9. Kirchner, T., Bannow, N., Grimm, C.: Analogue mixed signal simulation using spice and SystemC. In: Design, Automation Test in Europe Conference Exhibition, Nice, pp. 284–287 (2009)
10. Iskander, R., Louërat, M.-M., Kaiser, A.: Automatic DC operating point computation and design plan generation for analog IPs. Analog Integr. Circuit Signal Process. J. **56**, 93–105 (2008)
11. Iskander, R., Louërat, M.-M., Kaiser, A.: Hierarchical sizing and biasing of analog firm intellectual properties. Integr. VLSI J. **233**, 123–148 (2013)
12. Javid, F., Iskander, R., Louërat, M.-M.: Simulation-based hierarchical sizing and biasing of analog firm IPs. In: IEEE International Behavioral Modeling and Simulation Conference, San Jose, pp. 43–48 (2009)
13. Javid, F., Iskander, R., Durbin, F., Louërat, M.-M.: Analog circuits sizing using the fixed point iteration algorithm with transistor compact models. In: IEEE Mixed Design of Integrated Circuits and Systems, Warsaw, pp. 45–50 (2012)
14. Vachoux, A., Grimm, C., Einwich, K.: Extending SystemC to support mixed discrete-continuous system modeling and simulation. In: IEEE International Symposium on Circuits and Systems, Kobe, pp. 5166–5169 (2005)
15. Mu, Z., Van Leuken, R.: SystemC-AMS model of a dynamic large-scale satellite-based AIS-like network. In: Forum on Specification and Design Languages, Oldenburg, pp. 1–8 (2011)
16. Cenni, F., Scotti, S., Simeu, E.: Behavioral modeling of a CMOS video sensor platform using systemc AMS/TLM. In: Forum on Specification and Design Languages, Oldenburg, pp. 1–6 (2011)
17. Levi, T., Lewis, N., Tomas, J., Fouillat, P.: IP-based methodology for analog design flow: application on neuromorphic engineering. In: IEEE International NEWCAS-TAISA Conference, Montreal, pp. 343–346 (2008)
18. Saleh, R., Wilton, S., Mirabbasi, S., Hu, A., Greenstreet, M., Lemieux, G., Pande, P.P., Grecu, C., Ivanov, A.: System-on-chip: reuse and integration. Proc. IEEE **94**(6), 1050–1069 (2006)
19. Liu, W.: MOSFET Models for SPICE Simulation: Including BSIM3v3 and BSIM4. Wiley, New York (2001)
20. NXP. MOS model PSP level 103. http://www.nxp.com/models/mos/_models/psp/ (2011)
21. Enz, C., Krummenacher, F., Vittoz, E.: An analytical MOS transistor model valid in all regions of operation and dedicated to low-voltage and low-current applications. Analog Integr. Circuits Signal Process. J. **8**(1), 83–114 (1995)
22. Javid, F., Iskander, R., Louërat, M.-M., Dupuis, D.: Analog circuits sizing using bipartite graphs. In: IEEE International Midwest Symposium on Circuits and Systems, Seoul, pp. 1–4 (2011)
23. Libes, D.: Exploring Expect: A Tcl-Based Toolkit for Automating Interactive Programs. O'Reilly Media, Sebastopol (1994)
24. Maehne, T., Vachoux, A., Giroud, F., Contaldo, M.: A VHDL-AMS modeling methodology for top-down/bottom-up design of RF systems. In: Forum on Specification and Design Languages, Sophia Antipolis, pp. 1–7 (2009)

25. Graeb, H., Zizala, S., Eckmueller, J., Antreich, K.: The sizing rules method for analog integrated circuit design. In: IEEE/ACM International Conference on Computer Aided Design, San Jose, pp. 343–349 (2001)
26. Allen, P.E., Holberg, D.R.: CMOS Analog Circuit Design. Oxford University Press, Oxford (2002)

Chapter 7
Configurable Load Emulation Using FPGA and Power Amplifiers for Automotive Power ICs

Manuel Harrant, Thomas Nirmaier, Christoph Grimm, and Georg Pelz

Abstract In this paper we present a new concept of an application-oriented post-silicon verification method for automotive power micro-electronic devices. Automotive power semiconductors are mainly influenced by their real-life application but there is no sufficient method yet to assess device robustness within their application. For that reason we established a first approach to emulate different automotive power loads by running their model equations in real-time on an FPGA platform while the load current is controlled with a class AB power amplifier. The functionality of this approach is evaluated on the basis of automotive smart high-side switches and incandescent lamp models.

7.1 Introduction

Exploring and assessing robustness for automotive smart power micro-electronic devices become more and more difficult due to increasing complexity of the devices. This behavior is mainly driven by ever increasing customer demands for energy efficiency and safety. While there are well established methods for standard functional post-silicon verification [1, 2], there is less work done to assess device robustness within their real-life application. Application tests are done late during the verification process of automotive power devices.

While there is much freedom in accessing and changing all available operating conditions of the Device Under Test (DUT) and parameters of the power load

---

M. Harrant (✉) • T. Nirmaier • G. Pelz
Infineon Technologies AG, Am Campeon 1-12, 85579 Neubiberg, Germany
e-mail: Manuel.Harrant@infineon.com; Thomas.Nirmaier@infineon.com;
Georg.Pelz@infineon.com

C. Grimm
Technische Universität Kaiserslautern, Gottlieb-Daimler-Str., 67663 Kaiserslautern, Germany
e-mail: Grimm@cs.uni-kl.de

J. Haase (ed.), *Models, Methods, and Tools for Complex Chip Design*, Lecture Notes in Electrical Engineering 265, DOI 10.1007/978-3-319-01418-0_7,
© Springer International Publishing Switzerland 2014

during pre-silicon verification, application tests are using fixed setups to replicate the real system environment of the semiconductor. The usage of those standard loads (e.g. one specific DC motor for the wiper application in combination with the corresponding smart high-side switch) consists of a fixed set of load parameters and consequently give no statement about device robustness within their variations of all active and passive components as well as aging effects.

For that reason we want to explore the verification space by certain load parameters to achieve an application-oriented robust post-silicon verification approach. The concept is to emulate different automotive power loads like incandescent lamps, DC motors, lithium-ion batteries, etc. in real-time on an FPGA (Field Programmable Gate Array) and control the calculated load current with a class AB power amplifier circuitry. This approach allows to emulate application intrinsic parameters such as wire resistance and wire inductance of the connection between DUT and power load beside the load parameters itselves.

We present the first implementation of an experimental closed-loop test system at mid-range power performance to emulate different automotive power loads in real-time on an FPGA platform and explore the verification space by identified load parameters.

The outline of this chapter is set up as follows:

- Related work in the area of automotive power load emulation for lab verification topics is discussed in Sect. 7.2.
- Section 7.3 describes the experimental test setup regarding hardware concept and software architecture and its reached performance.
- The hierarchy for modeling automotive power loads in real-time for FPGA targets is explained in Sect. 7.4 and exemplified on the modeling process of an incandescent lamp.
- Section 7.5 compares an emulate incandescent lamp with nominal power of 21 W with the corresponding real lightbulb to evaluate the accuracy of the model while Sect. 7.6 show some selected measurement examples done for automotive front lighting.
- A conclusion and outlook regarding future work within this topic is presented in Sect. 7.7

7.2 Related Work

Several methods for load emulation exist with respect to Hardware-in-the-Loop simulation platforms [3, 4]. These platforms are almost used for testing integration of electronic control units (e.g. bus communication) or emulate their environmental conditions like sensor interfaces. But these solutions are not sufficient for real power load emulations which require current levels up to 90 A.

Other concepts [5–7] have sufficient time accuracy for the real-time emulation of automotive low-power loads, such as stepper motors or single battery cells, but are not satisfying for the emulation of high-power loads in multiple kW-ranges.

Similar to these low-power performance approaches there are also high-power emulation concepts [8–10, 12] available. These approaches can handle high load currents (up to 80 A) but are not able to handle the timing requirements which are needed to correctly replicate the dynamic behavior of DUT in combination with the power load.

A further limitation is the lack of flexibility in configuration. Automotive power micro-electronic devices go from simple single-channel power switches up to complex battery management devices which can handle several battery cells in parallel. For that reason, an efficient load emulation approach needs to be configurable for a multitude of devices that should be verified, based on lab measurements, in this way.

Combining the requirements for emulating automotive high-power loads during functional verification, the approach must fulfill the following tasks:

- Feedback test system required to handle interaction between DUT and power load.
- Real-time capability.
- Configurable number of load channels to handle single-channel as well as multi-channel devices.
- Output currents up to 90 A.

Consequently, no referenced concept meets the requirements to emulate automotive power loads for an application-oriented characterization method to the best knowledge of the authors.

7.3   First Experimental Setup

For emulating different automotive power loads in real-time we propose a closed-loop measurement test system consisting of an FPGA in combination with a class AB power amplifier. Due to strong interactions between micro-electronic devices and power loads within automotive applications an open-loop system, which cannot react to discontinuous dynamic behaviors (e.g. switching of power channels) is not sufficient for this topic.

Load emulation test systems, in general, consists of two parts:

- Any type signal processing unit (e.g. digital signal processors (DSP), microprocessors (μP), Field Programmable Gate Arrays (FPGA)) for evaluating the load model equations in real-time.
- A current source which controls and consumpts the load current depending on the model.

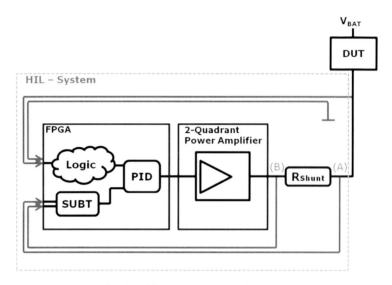

Fig. 7.1 Hardware concept for closed-loop load emulation for automotive power ICs

We have chosen an FPGA platform to process the physical equations of the load model in real-time. Referenced concepts are using digital signal processors (DSP) or microprocessors (μP) instead of FPGA platforms to solve the differential equations. Digital Signal Processors are often used because they are more arithmetic-oriented to solve even complex mathematical functions. However, reasons for choosing an FPGA are listed below:

- Low-level mathematical functions are solved faster (e.g. in few nanoseconds) due to high internal clock rates.
- Because of parallel operations and single-cycle timed loops numerous non-complex mathematical functions can be solved at the same time.

For instance, an FPGA platform seems to be the best trade-off between model complexity and processing speed for automotive power load models.

An overview to the first experimental test setup for real-time automotive power load emulation is presented in Fig. 7.1.

A reconfigurable load model (called "Logic" in Fig. 7.1) is running on the FPGA and calculating instantly the corresponding current values according to the DUT's output voltage. The power amplifier can work in two quadrants and is able to sink current for ohmic loads like incandescent lamps as well as additionally source current for inductive/capacitive loads like DC Motors, batteries, etc. Furthermore, with respect to the used power amplifier concept (see a simplified schematic in Fig. 7.2a), a digital PID controller and a mathematical subtraction (SUBT) are implemented inside the FPGA because of single-ended ADC channels.

The implemented power amplifier is a controlled voltage source that consists of parallel driven complementary power transistors with high collector currents and

7 Automotive Power Load Emulation

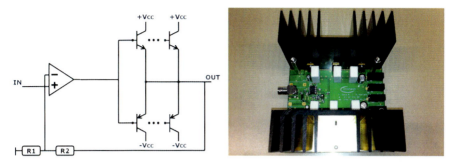

Fig. 7.2 First approach of a class AB power amplifier for automotive power load emulation

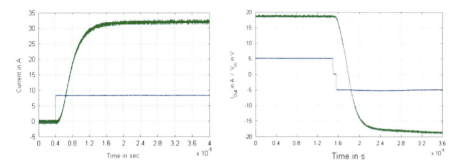

Fig. 7.3 Performance evaluation of the class AB power amplifier

an operational amplifier. The module itself adjusts the potential at (B). The digital PID controller, implemented on the FPGA, controls the load current. This is done by using the calculated load current as reference and the voltage between (A) and (B) as process variable. Current consumption is consequently defined as the voltage drop at a shunt resistor.

This combination of analog class AB power amplifier circuitry and digitally implemented PID controller acts as a controlled dynamic current source and is operating in two quadrants in the complete experimental setup. The reason of choosing this type of power stage is the fast closed-loop speed and the good handling of interactions between DUT and power load as well as other discontinuous behaviors, like switching of power channels, during operation.

Due to the high current consumption when emulating automotive power loads, many challenges needed to be solved, such as impedance matching and symmetry resistors in the output path to prevent thermal drift of the power transistors or several solutions to get a stable as well as fast control loop.

Figure 7.3 shows the step responses for a first approach of the power amplifier in order to evaluate its performance regarding speed and power. The module's input was driven with an arbitrary waveform generator (AWG) to produce the stimuli

Table 7.1 Large signal performance of the closed-loop system

Parameter	Performance range
V_{Bat}	0 V ... 24 V
I_{OUT}	−40 A ... +40 A
t_{FPGA}	6 µs ... 10 µs
$t_{PowerStage}$	8 µs

while the output current was measured at an 0.68 Ω resistor against ground. Due to the used operational amplifier's loop-gain, the output current is higher than the input step depending on the values of R1 and R2.

First results of the class AB amplifier's large signal performance can be summarized as follows:

- Maximal output current of 40 A measured in combination with the first approach of the closed-loop system.
- The step from 0 to 32 A has a settling time of approximately 8 µs.
- The module has a reaction time between input signal and the rising slope of the output current.
- Transition between operating quadrants (in this case a change from first to third quadrant was chosen) occurs in less than 8 µs (Table 7.1).

Performance limitations of the complete measurement setup come from several points of the closed-loop system. On the one hand side, the digital-to-analog conversion unit of the FPGA has a maximal sampling period of 10 µs, dependent on the number of used conversion channels. Secondly, the operational amplifier used for controlling the potential at (B) is limited because of its finite slew rate and operating frequency. Summarizing the results coming from the presented performance evaluation lead to the following specification of the closed-loop test system:

This experimental setup can be controlled and configured using a stand-alone software programmed in LabVIEW for rapid prototyping purposes. The graphical user interface (GUI), see Fig. 7.4, allows the configuration of available loads including their individual parameters, trigger measurements, monitor selected waveforms of interest and brings the test system in a safe operating state in case of malfunctions.

Furthermore, the test system can be connected to a PXIe (PCI eXtension for Instrumentation) System and fully controlled via a lab automation environment [11]. The automation is controlling all required equipment via General Purpose Interface Bus (GPIB), PXI bus and allows to change all identified load parameters on-the-fly. This approach allows to run variations up to large statistical Monte Carlo experiments in an automated way including an automated report generation.

7 Automotive Power Load Emulation

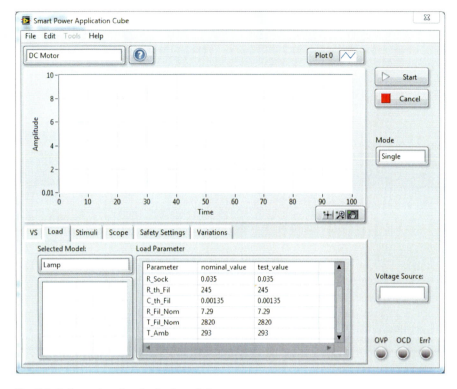

Fig. 7.4 Software interface for load emulation

7.4 Load Modelling for Real-Time Evaluation

As already mentioned, there are different solutions for digital signal processing units to calculate automotive power load models in real-time. The main option lies on DSP and FPGA solutions. There are automated ways of code generation from MATLAB/Simulink models to DSP and μP targets but no automated flow could be found for FPGAs as target platform. An overview to emulator development options is visualized in Fig. 7.5.

The modelling process starts with the analytical equations of the power load. Load models are commonly available in MATLAB/Simulink, VHDL-AMS or SystemC-AMS. Especially AMS languages differ in abstraction level, which makes it more or less difficult to transfer the models to another abstraction. MATLAB/Simulink is the standard approach for automotive system level modelling, while VHDL-AMS is standard for block level.

Mixed-Signal models are often written as implicit differential equations and simulators require implicit solvers. Implementing such solvers on an FPGA platform will need lots of resources and never meet the required processing speed

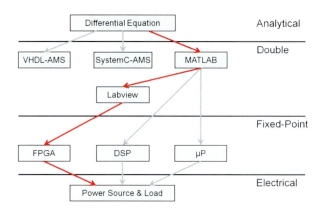

Fig. 7.5 Hierarchy for load emulation development

to emulate automotive power loads. For that reason, we have choosen LabVIEW (translated from MATLAB/Simulink models) and a semi-automated way to get from the analytical differential equations to a synthesized and executable digital load model that can be implemented on an FPGA without the need of any implicit solver algorithm. The advantages of this flow are listed below:

- The graphical interface of LabVIEW's FPGA toolbox is similar to MATLAB/Simulink and allows an easy transfer of block level Simulink models.
- The FPGA library from LabVIEW is a subset of the MATLAB/Simulink's library.
- An integrated toolbox allows an automated way to get a digital FPGA design out of the block schematic.
- LabVIEW is a common environment used for lab automation, which enables easy access to FPGA configuration as well as other lab equipment needed for load emulation topics.
- FPGA platforms can easily be integrated in automated measurement and test systems running with LabVIEW software.
- Optimization and debug within the block level model can already be done in MATLAB/Simulink to save development time (no need for synthesis of VHDL code).

The procedure for real-time load emulation using LabVIEW in combination with an FPGA platform is done in three steps. They are described below based on the example of incandescent lamps:

1. **Physical equations:**
   Starting first with Kirchhoff's law for electrical meshes (see Eq. (7.1)). This equation considers the influence of wire resistance as well as wire inductance beside the thermo-electrical model of the lamp on the load current $i(t)$.

7 Automotive Power Load Emulation

Kirchhoff Voltage Law:

$$u_{HS}(t) = R_{Wire} \cdot i(t) + L_{Wire} \cdot \frac{di}{dt} + R(T) \cdot i(t) \tag{7.1}$$

where u_{HS} is the time-variant output voltage of the smart high-side switch, R_{Wire} and L_{Wire} are resistance and inductance of the conductor and $R(T)$ is the thermal-dependent resistance of the filament.

In a second equation the load specific behavior, depending on electro-thermal or electro-mechanic loads must be included. This behavior might be thermal heating when modelling incandescent lamps, back electro-magnetic force (Back-EMF) for all kinds of motors, etc.

Energy Conservation:

$$P_{el} = P_{rad} + P_{cond} + C_{th} \cdot \dot{T}_{Fil} \tag{7.2}$$

$$\dot{T}_{Fil} = \frac{1}{C_{th,Fil}} \cdot (R(T) \cdot i^2(t) - \gamma \cdot (T_{Fil} - T_{Amb})^4 - \frac{T_{Fil}}{R_{th,Fil}}) \tag{7.3}$$

Beside Kirchhoff's law the energy balance in combination with the thermal equivalent circuit of the lamp is used to consider thermal heating of the lamp's filament, which effects an increasing resistance. The energy balance simply says that the electrical power P_{el}, which is dependent on the thermal-variant resistance, can be split into radiated power P_{rad} (see Stefan Boltzmann's law for black body radiation), conducted power P_{cond} and thermal heating.

Thermal resistor $R(T)$:

$$R(T) = R_{Fil,nom} \cdot (\frac{T_{Fil}}{T_{Fil,nom}})^p \tag{7.4}$$

The third equation calculates the thermal-variant resistance of the filament on the basis of its resistance $R_{Fil,nom}$ and filament temperature $T_{Fil,nom}$ at nominal power.

2. **FPGA Implementation:**
Implementing Eqs. (7.1), (7.3) and (7.4) on an FPGA platform leads to the following block level design, which is afterwards converted to VHDL and synthesized within an automated way (Fig. 7.6).

Equations (7.1), (7.3) and (7.4) were transformed from differential equations to a set of difference equations first order that must be calculated in real-time on the digital signal processing unit. The sampling time t_{Sample} is equal to the processing time of one loop cycle. This timing strongly depends on the used FPGA and analog-to-digital converter as well as the model complexity itself. For the presented lamp model, the sampling time was $\approx 7\,\mu s$. Factors or parts of the

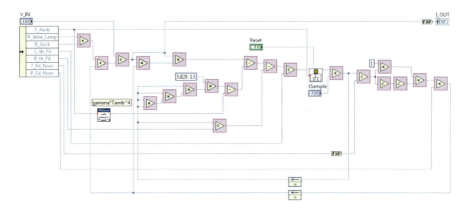

Fig. 7.6 Real-time capable incandescent lamp model in LabVIEW's FPGA toolbox

equation which are time-invariant ($\gamma \cdot T_{Amb}^4$) are calculated in a pre-processing step to optimize the digital design. Doing this pre-processing step leads to two advantages regarding performance of the model implementation:

- The number of mathematical operations inside the FPGA can be reduced which leads to a faster processing speed.
- The model can be optimized regarding the number of bits spent for fixed-point calculations which also leads to a short calculation time and less resources of logic blocks.

3. **Parameter extraction:**
   In a last step the influencing factors of the lamp model are figured out. There are several sensitivity analysis methods to identify these parameters such as evaluating the partial derivative for every factor to see the impact of each parameter to the model's output. This is one of the most reasonable advantages of emulating automotive power loads instead of using the equivalent real load.

   An overview to these parameters, including their range or nominal value, for an incandescent lamps with 21 W nominal power can be found in the following Table 7.2:

   In contrast to the power load, where all parameters are nominal values for the specific type of lightbulb, the conductor parameters and the ambient temperature are given within ranges. Conductor parameters depend on the implementation inside the vehicle as well as on the location of the load compared to the electronic control unit.

Similar to the steps that have been made for incandescent lamps it is possible to generate parameterized real-time models for several types of automotive power loads, such as motors, LED modules, batteries or even full applications like electronic throttle control (motor and throttle) that are application-relevant for automotive power micro-electronic devices and execute them on an FPGA platform.

Table 7.2 Influencing parameters of the incandescent lamp model with 21 W nominal power consumption

Conductor	Description	Range
R_{Wire}	Resistance of the conductor	$30\,m\Omega \ldots 105\,m\Omega$
L_{Wire}	Inductance of the conductor	$1.5\,\mu H \ldots 4.5\,\mu H$
Lamp factors	Description	Nominal value
$C_{th,Fil}$	Heat capacity of lamp's filament	$13.5\,\frac{mJ}{W}$
$R_{th,Fil}$	Thermal resistance of lamp's filament	$245\,\frac{K}{W}$
$T_{Fil,nom}$	Filament temperature at nominal power	$2{,}820\,K$
$R_{Fil,nom}$	Filament resistance at nominal power	$7.3\,\Omega$
R_{Sock}	Resistance of the lamp's socket	$0.0035\,\Omega$
Environment	Description	Range
T_{Amb}	Ambient temperature	$-40\,°C \ldots +150\,°C$

Furthermore, the emulation test system can be configured to handle dynamic load steps for power loads that cannot be modelled on the basis of physical equations, such as micro-controllers or xenon lighting profiles.

7.5 Evaluation of Lamp Model

This approach of a closed-loop load emulation system was tested by emulating an incandescent lamp for automotive lighting applications, such as a direction indicator light. The transfered lamp model is running in real-time on the FPGA platform while the power amplifier generates and controls the calculated load current. Incandescent lamps are non-linear ohmic loads and the amplifier was only operating in the third quadrant and sink the load current which is consumed by the lamp.

The evaluation of the lamp model should have the following purpose:

- Evaluate the dynamic performance of the closed-loop load emulation test system, for continuous as well as discontinuous behavior.
- Evaluate the precision of the lamp model to see whether the incandescent lamp was correctly characterized (Fig. 7.7).

Running several measurements and easily compare results from the closed-loop test system with an equivalent real incandescent lamp, the following testbench was used:

The used DUT (A) for this evaluation was a single-channel smart high-side switch which is commonly used for front lighting applications. In the area of automotive power micro-electronic devices, so called "Smart Power ICs" include several protective functions (e.g. over-current and over-voltage detection, over-temperature shut-down, current sense capability, etc.) to prevent its circuitry from

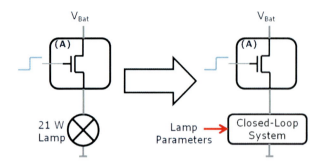

Fig. 7.7 Smart high-side switch with real 21 W lamp (*left*) and emulated 21 W lamp (*right*)

damages. In a first series of measurements the smart high-side switch was operating with a real incandescent lamp at its output. Similar during a second series of measurements, the closed-loop system was connected to the same smart high-side switch. Afterwards the waveforms of real systems were compared to the emulated ones.

The first evaluation of results (see Fig. 7.8) was focused on the accuracy of the lamp model during continuous behavior. Using the same device in both series of measurement prevent variations caused by different samples (e.g. manufacturing tolerances according to wafer position). In general, incandescent lamps have high inrush currents (approximately higher by factor 10 than its steady state current) due to their low resistance at low temperatures. When current is flowing through the filament the resistance is increasing in a non-linear way and the current is decreasing inversely.

Remaining deviations between emulated current waveform and the behavior of the real incandescent lamp may have several reasons, such as:

- Finite accuracy from fixed-point arithmetic blocks for the FPGA implementation and resolution of the ADC/DAC modules.
- Quantization and Sample and Hold characteristic of the analog-to-digital converter module.
- Latencies coming from the mixed-signal part between converting the output voltage of the smart high-side switch into discrete values and the control of the load current using the power amplifier.

Second evaluation results presented in this chapter concentrate more on the hardware performance of the closed-loop test system. In real-life applications the lamp's inrush current exceeds the value of the smart switches current limitation. As far as the current consumption of the load exceeds this specified value I_{Lim} the switch shuts-down for a certain period of time to avoid power drop or thermal destruction until it is automatically switched on again. The procedure is repeated as long as the current consumption stays above I_{Lim}. This discontinuous behavior is called thermal toggling and occurs every time lamps are switched on inside automotive lighting

7 Automotive Power Load Emulation

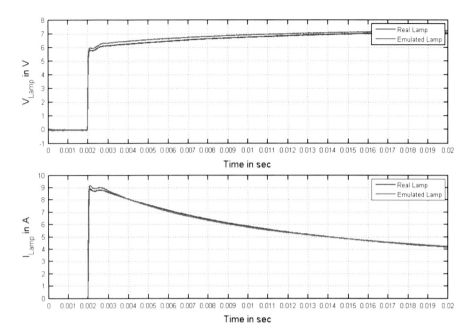

Fig. 7.8 Comparison of real lamp (*blue*) and emulated lamp (*green*) in continuous behavior

applications. The manner how the closed-loop system handles this toggling process is shown in Fig. 7.9.

When the smart high-side switch is turned on, current consumption is defined through the voltage waveform and the model equations. As far as the current exceeds I_{Lim} the device starts toggling until it can handle the load current of the real/emulated lamp. According to Fig. 7.9 it is clear that the closed-loop system can handle this discontinuous behavior and correctly reproduce the current waveform with only small deviations. This discontinuous behavior is somehow thermal-dependent and the point in time cannot be predicted. This discontinuous toggling is of great interest during post-silicon verification of smart high-side switches and was sometimes limiting the functionality when using electronic load equipment for emulation tasks.

In conclusion, the results achieved during this evaluation measurements in Figs. 7.8 and 7.9 must be evaluated. This analysis is shown in Fig. 7.10.

Due to strong interactions between the device under test and automotive power load in combination with non-reproducible results (caused by noise, timing jitter and dynamic behavior of the complete system) the mean value of 10 measurements was used as a precision scale.

To evaluate the model's precision the percentage derivation between real current waveform and emulated current waveform was calculated. It can be stated that the relative error over time never exceeds a 5 % interval for the emulated incandescent lamp. The maximal derivation seems to be at the rising slope of the lamp's inrush current. The reason for this behavior could either be caused by the model calculation

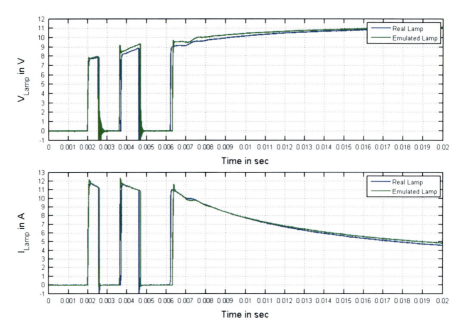

Fig. 7.9 Comparison of real lamp (*blue*) and emulated lamp (*green*) in discontinuous behavior

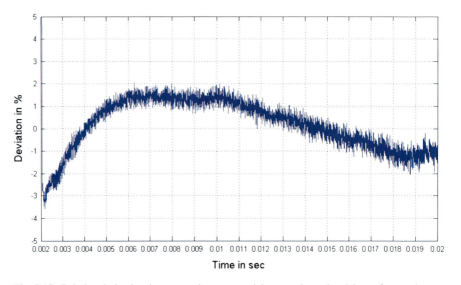

Fig. 7.10 Relative derivation in current between real lamp and emulated lamp for continuous behavior

(numerical noise, accuracy of AD/DA converters) or by the dynamic behavior of the closed-loop system itself. Improving the system's closed-loop speed could minimize this issue and achieve a even better precision.

7.6 Experimental Results

The great advantage of using such an emulation test system instead of using physical automotive power loads is the access to all load parameters that may influence the dynamic behavior or even create situations in which the device under test is not working properly anymore. In a first run we have chosen sweep tests and statistical Monte Carlo experiments to visualize the impact of operating conditions and parameter spread from the modelled incandescent lamp model.

First measurements, again, show the real-life behavior how incandescent lamps are switched on inside automotive applications, as already described in the second part of Sect. 7.5. Since we know that the filament temperature is correlated to the ambient temperature when the lamp is in off-state as well as to the resistance, we performed a sweep test to see the impact of variable ambient temperature on the shape of the current waveform.

It is obvious, according to the results presented in Fig. 7.11, that the ambient temperature of the lamp will have an impact on the maximum inrush current and, which is more interesting for the micro-electronic device, on the dynamic behavior between high-side switch and incandescent lamp. As far as the operating temperature lasts from $-40\,°C$ to $+150\,°C$ it is necessary to check the behavior of this application under different operating conditions. Measurements on the physical light bulb will take a long time caused by cool-down phases of the lamp to get back to its off-state resistance. Here in this case we can see the thermal toggling effect caused by the over-load protection of the smart high-side switch as far as the ambient temperature of the light bulb was adjusted below 293 K. Performance evaluations in this case could be to measure the length of the toggling pulse until the smart switch is turned on again or the number of toggling pulses.

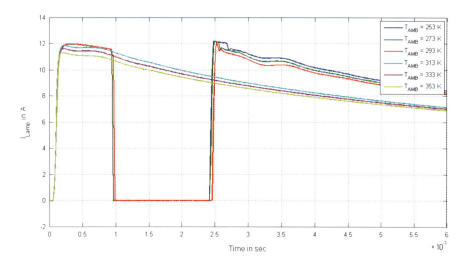

Fig. 7.11 Current consumption for incandescent lamp with 21 W nominal power at different ambient temperatures

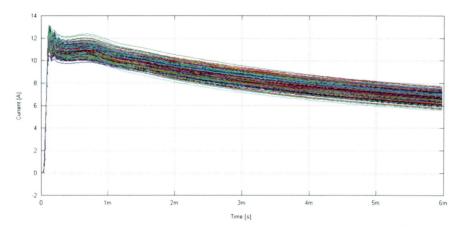

Fig. 7.12 Current waveform for 1,000 Monte Carlo experiments and Gaussian distributed incandescent lamp model parameters

Second measurement runs shall give more feedback regarding device robustness and robustness metrics. There is a multitude of measures and metrics with statistical background in semiconductor industry [15] to get a relationship between performance distribution and its specification as presented in [13]. For evaluating our results we have chosen the Worst-Case-Distance (WCD), which was proposed in [14] for yield and design analysis, for the current limitation of the selected smart high-side switch.

The first step was to perform a statistical Monte Carlo experiment with 1,000 runs. In fact that a specific incandescent lamp will have a nominal set of parameters including a spread coming from manufacturing tolerances a Gaussian distribution was carried out for statistical Monte Carlo experiments. The standard deviation σ for these parameters was set to 5% from its nominal value. The output of interest from these measurements was the maximum value of the inrush current dependent on the lamp's parameters.

Figure 7.12 shows the current waveform of 1,000 Monte Carlo experiments within one single plot to see the deviation of the current shape dependent on all lamp parameters. It is obvious that there is a large spread in current consumption even for small spread of load parameters.

Getting a little deeper into robustness measures and robustness metrics we will have a closer look on the maximum inrush current with respect to the lamp's parameters. Visualizing correlations between either maximum inrush current and lamp parameters or the lamp parameter itselves is done using Matrix plots. According to Fig. 7.13 gaussian distributed lamp parameters lead to a distribution of the maximal inrush current with a nominal value of approximately 12 A. Imagine the current limitation of the used smart high-side switch would be specified as $I_{LIM} = 14$ A we can calculate a Worst-Case Distance as described in [14]:

7 Automotive Power Load Emulation 125

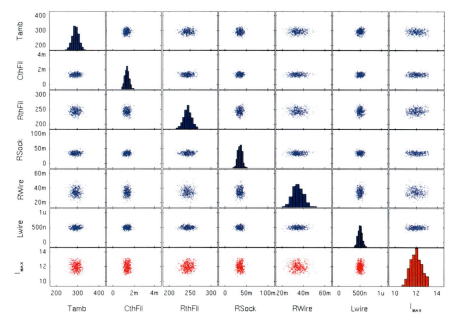

Fig. 7.13 Matrix plot of 1,000 Monte Carlo experiments for maximum inrush current (I_{max}) related to lamp parameters

$$WCD = 7.8435 \qquad (7.5)$$

This value gives a correlation between nominal value of the maximum inrush current and the current limitation as specified for the given high-side switch in multiples of the standard deviation σ.

7.7 Conclusion and Outlook

In this chapter we presented a flow to emulate different automotive power loads in real-time on an FPGA platform and control the required load current using a class AB power amplifier circuitry. First results have shown that this concept is working with a closed-loop speed that meets the requirements for emulating mid-power loads and open a way to assess the robustness of automotive power microelectronic devices with respect to their real-life application already during post-silicon verification.

For future load emulation it is necessary to improve the performance of the existing closed-loop system for high-power loads, which have a current consumption up to 90 A, as well as extend the approach by parallelized channels to emulate loads for multi-channel devices, especially in the area of battery management.

Achieving more power goes hand-in-hand with a reduction of the closed-loop speed. For that reason, the control loop must be optimized (e.g. using a pre-amplifier to handle increasing capacities at the output of the operational amplifier).

Besides improving the performance of the closed-loop system further automotive loads must be modelled in this way and implemented on the FPGA. The emulation of lithium-ion cells for lab verification topics of battery management devices stands in the foreground especially for future topics regarding electro-mobility.

References

1. Nirmaier, T., Meyer zu Bexten, V., Tristl, M., Harrant, M., Kunze, M., Rafaila, M., Pelz, G., Lau, J.: Measuring and improving the robustness of automotive smart power microelectronics. In: Design, Automation and Test in Europe, Dresden, pp. 872–873 (2012)
2. Nirmaier, T., Harrant, M., Pelz, G.: Extending constrained random verification to mixed-signal automotive power devices using a non-stationary Markov process. In: International Workshop on Silicon Debug and Diagnosis, ITC, Anaheim (2011)
3. Duelks, R., Salewski, F., Kowalewski, S.: A real-time test and simulation environment based on standard FPGA hardware. In: TAIC part, Windsor, pp. 197–204 (2009)
4. Dufour, C., Belanger, J., Lapointe, V.: FPGA-based ultra-low latency HIL fault testing of a permanent magnet motor drive using RT-LAB-XSG. In: POWERCON, New Delhi, pp. 1–7 (2008)
5. Thanheiser, A., Kohler, T., Herzog, H.-G.: Battery emulation considering thermal behavior. In: Vehicle Power and Propulsion Conference (VPPC), Chicago, pp. 1–5 (2011)
6. Thanheiser, A., Meyer, W., Herzog, H.-G.: Design and investigation of a modular battery simulator system. In: Vehicle Power and Propulsion Conference (VPPC), Dearborn, pp. 1525–1528 (2009)
7. Srinivasa Rao, Y., Chandorkar, M.C.: Real-time electrical load emulator using optimal feedback control technique. In: IEEE Transactions on Industrial Electronics, Kandy, vol. 57, pp. 1217–1225 (2010)
8. Ginot, N., Le Claire, J.C., Loron, L.: Active loads for hardware in the loop emulation of electro-technical bodies. In: IEEE Industrial Electronics Society, Raleigh (2005)
9. Rao, Y.S., Chandorkar, M.: Electrical load emulation using power electronic converters. In: IEEE Region 10 Conference, Hyderabad (2008)
10. Armstrong, M., Atkinson, D.J., Jack, A.G., Turner, S.: Power system emulation using a real-time 145 kW virtual power system. In: IEEE European Conference on Power Electronics and Applications, Dresden (2005)
11. Kunze, M., Pirker-Fruehauf, A.: A novel methodology to combine and speed-up the verification process of simulation and measurement of integrated circuits. In: AUTOTESTCON, Salt Lake City, pp. 259–262 (2008)
12. Grubic, S., Amlang, B., Schumacher, W., Wenzel, A.: A high-performance electronic hardware-in-the-loop drive-load simulation using a linear inverter (LinVerter). In: IEEE Transactions on Industrial Electronics, Kandy, vol. 54, pp. 1208–1216 (2010)
13. Nirmaier, T., Kirscher, J., Maksut, Z., Harrant, M., Rafaila, M., Pelz, G.: Robustness metrics for automotive power microelectronics. In: Design, Automation and Test in Europe, RIIF Workshop, Dresden (2013)
14. Antreich, K., Graeb, H., Wieder, C.: Circuit analysis and optimization driven by worst-case distances. In: IEEE Transactions on Computer-Aided Design of Integrated Circuits and Systems, San Jose, pp. 57–71 (1994)
15. NIST/SEMATECH e-Handbook of Statistical Methods, http://www.itl.nist.gov/div898/handbook/

Chapter 8
Model Based Design of Distributed Embedded Cyber Physical Systems

Javier Moreno Molina, Markus Damm, Jan Haase, Edgar Holleis, and Christoph Grimm

Abstract In this chapter, we propose a Model-Based Design (MBD) methodology that aims to deal with complexity due to the convergence of different domains and technologies in distributed embedded systems, enabling early design optimization and reduction of time-to-market. This methodology requires models for very different domains able to work together: Electronic System Level (ESL), network, radio propagation and quantities to be captured by the sensor systems. Using different simulators involves co-simulation and coupling overhead. We introduce a framework based exclusively in SystemC and its extensions for Transaction-Level Modeling (TLM) and Analog Mixed-Signal (AMS), and extensible with additional C/C++ code. The whole approach has been validated in a Cyber-Physical System for demand side energy management in buildings and environments, developed during the SmartCoDe Project.

J.M. Molina (✉) • M. Damm • C. Grimm
Technische Universität Kaiserslautern, Gottlieb-Daimler-Str.,
67653 Kaiserslautern, Germany
e-mail: moreno@cs.uni-kl.de; damm@cs.uni-kl.de; grimm@cs.uni-kl.de

J. Haase
Vienna University of Technology, Institute of Computer Technology,
Gußhausstr. 27-29/384, A-1040 Vienna, Austria
e-mail: haase@ict.tuwien.ac.at

E. Holleis
Tridonic, Dornbirn, Austria
e-mail: edgar.holleis@tridonic.com

J. Haase (ed.), *Models, Methods, and Tools for Complex Chip Design*, Lecture Notes in Electrical Engineering 265, DOI 10.1007/978-3-319-01418-0_8,
© Springer International Publishing Switzerland 2014

8.1 Introduction

Over the last years, electronic systems have become more and more complex and design tools and approaches have had to be adapted accordingly. System-on-Chips (SoCs) are now heterogeneous analog mixed-signal devices with both digital and analog parts, which makes the design process more complicated. This complexity has increased even more with new application paradigms, such as Wireless Sensor Networks and some Cyber-Physical Systems.

For instance, SmartCoDe project investigated the development of a Cyber-Physical System to control and manage energy demand in buildings and environments in order to adapt it to the energy generated by volatile energy sources. This area of research is very active because most of the available renewable energy sources have this volatile behaviour. Energy generation cannot be planned and controlled and is scattered geographically, on contrast with traditional power plants. The approach in SmartCoDe consists in a distributed cyber-physical system, integrated by embedded systems attached to energy suppliers and energy consumers and an energy management application which analyses the available information about energy supply (e.g. wind forecast) and controls the functioning of the different energy consumers according to it and some policies which vary depending on the kind of appliance.

Designing such a system involves many complex design decisions which condition all the following steps. Requirements for embedded systems vary depending on the distribution of computational resources within the system. This distribution has to take into account limitations of all hardware, software and communication. Furthermore, some limitations might even depend on where the embedded systems are installed, due to the environment interaction. Figure 8.1 depicts the complexity of this kind of systems.

8.1.1 Model-Based Design Approach

In embedded systems design, Model-Based Design methodologies have provided means to deal with this complexity and today it is a very common design approach. Nonetheless, the new additional degrees of complexity have not been successfully integrated in the Model-Based Design approach, mainly because combining so many different models into an integrated multidisciplinary model is a very challenging task.

Thus, on the one hand, the use of sensors and actuators introduces interaction with the environment. Therefore, in order to predict the system behaviour, it is necessary to take into account the temporal variation of the quantities to be sensed. In a monitoring system, for instance, where data is transmitted only when certain difference is detected, volatile measures require more transmissions and more energy consumption, which is a major constraint in these systems that has to be carefully assessed at design time.

8 Model Based Design of Distributed Embedded Cyber Physical Systems

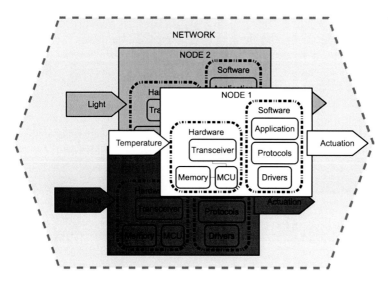

Fig. 8.1 Complexity in cyber-physical systems

However, modelling the physics significantly increases complexity of the simulation. Continuous variation must be included in a discrete event simulation. This involves mixing and synchronizing different Models of Computation (MoCs), which typically means coupling different simulators.

On the other hand, WSNs and similar distributed systems have to be modelled as a whole. Isolated node models are insufficient to validate and verify the system or to evaluate node hardware requirements, as they are affected by the distribution of computational resources, the network topology and the network traffic. A bad network topology or an inappropriate routing algorithm might overload some nodes, with the subsequent effect in system reliability. Therefore, network and propagation have to be considered in order to achieve successful designs.

To model the effects of distribution and network, not only one node has to be modelled, but several nodes (possibly thousands) together with the interaction among them. This requires a higher level of abstraction to still get acceptable simulation performance.

8.1.2 Multi-domain Simulation

In order to deal with all this different models, the first approach is to create ESL, physics and network models separately using already available specific tools. If results are evaluated independently, the interaction between them will be missing and unpredicted situations may arise during later design stages. This not only may lead to critical problems and potential risks, but also makes optimization more difficult. Interaction between the models can be achieved by co-simulation. However, co-simulation requires the different simulation environments to be running in parallel and to frequently synchronize among them, affecting significantly the overall simulation performance.

Hence, the idea of a single multi-level and multi-domain simulation environment arises. Such a solution would provide better model integration, which results in better performance and better and more intuitive knowledge of the whole system. Nevertheless, there is still no complete and satisfactory solution.

In this work, a unified simulation framework which comprehends ESL, analog, network and propagation models, is presented. This unified simulation framework has been used to model a WSN to control the energy demand of appliances in a building. A sensor node is attached to every device and depending on the environmental conditions and the renewable energy availability, decisions are made and the nodes can actuate on the devices they are attached to and set the most suitable states.

The rest of the paper is organised as follows: After discussing some related work, we make some notes on the methodology behind our simulation approach. In Sect. 8.4, the implementation of the WSN simulation is described. Section 8.5 describes the energy management application which is developed with the help of the simulation framework and shows some simulation results before concluding.

8.2 Previous Work

Already with VLSI systems the need of dealing with different levels of abstraction and different domains became a focus of research in electronics. In this context, in 1983, Gajski et al. introduced the Y-Chart which represents the interconnection between different levels of abstraction in three different planes: structural, behavioural and physical [9].

In embedded systems design, apart from the increasing complexity of the SoCs, the very demanding requirements, for instance in terms of efficiency and power consumption, have led to specific operating systems and software which had to be developed in parallel to the hardware. This interconnection between hardware and software design motivated the appearance of new design methodologies, such as Electronic System Level (ESL) and Model Based Design (MBD) [4].

Hence, the new methodologies and abstraction levels required to design networked embedded systems and interaction with the environment, which are significant challenges in Wireless Sensor Networks (WSN) and Networked Cyber Physical Systems (NCPS) design [13, 17], are just another step further in electronic systems design.

There are already some Model-Based Design approaches in order to consider physical processes at design time. In [11], a MBD methodology for Cyber-Physical Systems is proposed. This methodology consists of ten steps, in which physical models are already used for problem characterization. However, this methodology is focused on the coupling of embedded systems with physical environment interaction, but no attention is paid to system distribution.

The main obstacle for using MBD is the lack of appropriate tools. In [3], a Modelica based physical model is combined with the ns-2 network simulator. Special attention has to be paid to synchronization. Nevertheless, it is not an ESL tool, where node hardware and software could be easily and precisely modelled, and therefore it is not well suited for system design.

Ptolemy II [16] enables simulation of different levels of abstraction and domains through different Models of Computation (MoCs) within the same simulator.

SystemC is an industry standard for ESL design. As it is based on C/C++ it is easily extensible and there are already some SystemC-based simulators capable of modelling networks, such as those presented in [5] and [7]. The simulator in [5] leverages Transaction Level Modeling (TLM) [1] to improve simulation performance and to abstract energy consumption information [15]. It is even possible to include in the simulation Instruction Set Simulators (ISS) to obtain cycle-accurate microcontroller models [18].

Although SystemC provides discrete-event simulation, using SystemC-AMS extensions, it is also possible to model continuous time processes. Therefore, ESL (SystemC), network (TLM based wireless model) and physical processes (SystemC-AMS) can be all modelled together within the same SystemC simulation.

Moreover, as SystemC is C/C++ based, software models can be reused as real application code, as long as some coding rules are followed, such as, using C based coding style and avoiding dependencies from libraries that will not be available in the embedded system.

During the work presented in this chapter continues the work realized in [5, 15] and [18], by enhancing and extending the simulation environment, including physical processes to model environment interaction. This physical processes are implemented in SystemC-AMS [2].

Furthermore, this chapter provides a new use case for smart energy management and building automation, where the design methodology and simulation approach have been tested.

8.3 Methodology

Distributed embedded systems are the convergence of several technologies and have very demanding requirements such as reduced size of the embedded systems, low cost and high efficiency, specially in terms of power and energy. The design process must include cross-level optimization in order to fulfil those requirements.

Modelling each aspect separately not only leads to insufficient optimization, but also involves a risk, due to unexpected effects of interactions among the different system domains. Furthermore, in distributed cyber-physical systems, interactions with external environment have to be considered as well.

Therefore, ordinary hardware design methodologies or even embedded systems hardware and software co-design methodologies might not provide fully satisfactory results. One of the main challenges in CPS is to find an appropriate methodology that could lead to optimum results.

Simulation plays a crucial role in CPS design. Apart from the advantages of simulation in ordinary embedded systems design, CPS operation depends on environmental conditions which are imposed by the location where they are installed or deployed. These conditions might be difficult, if not impossible, to recreate in a testbed. Furthermore, a significant part of the applications require them to run unattended for very long periods of time. Simulation enables the evaluation of both environmental conditions and long term operation.

8.3.1 Requirements

The first step is the requirements elicitation of the system to design. Based on these requirements, a network architecture has to be defined, identifying the roles of the different components and, accordingly, the requirements for the embedded systems that will integrate the system. This crucial step rules the whole methodology. Fast exploration and refinement is crucial in order to achieve a successful design. Network models and the physical models of the quantities sensed by the sensors are of great aid in order to estimate nodes and network activity and correctly dimension the system. Upon refinement of the distribution architecture, embedded systems requirements may vary significantly, and, therefore, most important decisions have to be made before any specific knowledge about the hardware/software platforms is available, but considering their feasibility.

8.3.2 Functional Model

Once the requirements of the hardware/software platforms are set, their basic logical architecture is defined and the interfaces specified. The next goal is to obtain a purely functional and comprehensive executable specification of the system. This high level model will be the basis of the whole development. The functional model must start as generic as possible. While the system components are fixed, models can be refined and more detailed and specific models can be implemented if necessary. As depending on the application requirements, criticality of the different elements may vary, the model must be flexible enough and support real multi-level simulation, implementing low-level models only in the specific components that require it, so that the trade-off between accuracy and performance is correctly handled.

One of the main milestones in this methodology is the integration of the functional model within the network and environment models. This step enables refinement of the distributed model just before hardware/software partitioning starts and changes are still affordable.

8.3.3 Hardware/Software Co-design

At this point, the model must assist in evaluating the best SoC partitioning, as it is common practice in model-based design methodologies for embedded systems. Hardware and software must then be designed in parallel and simulated models provide the necessary cross-layer knowledge to hardware and software designers to create a consistent and optimal design.

At the same time, implementation of the models can be refined into more specific models, that can be used as a virtual prototype, where the system and the application can be validated and verified. This enables error detection, debugging and therefore reducing the risks of critical failures in implementation. When the virtual prototype is sufficiently mature, the real prototype can be manufactured.

Last, but not least, virtual prototypes can also be used to verify the resulting hardware/software platform.

8.3.4 Deployment

When the hardware prototype is available, the virtual prototype can be calibrated in order to obtain more accurate and realistic results. This is another critical step which is characteristic of distributed embedded systems design, where the system deployment is as important as the platform manufacturing process.

In many cases, once deployed, the only way to access or modify the systems will be through software over-the-air updates and no further maintenance options will be feasible. Moreover, the results of a single node prototype or a small network of prototypes can be used to calibrate and refine the whole networked system model and evaluate weaknesses and long-term performance. Therefore, evaluating the complete system with the most realistic data might solve some problems that later will be very difficult to address.

Figure 8.2 summarizes the whole methodology explained in this section.

8.4 Models Implementation

This section describes the implementation of the models required for the design of the energy demand management CPS introduced in Sect. 8.1. The main purpose of the implementation is to comprehend all multi-domain models in the same simulation environment. The approach followed is to use SystemC as the simulator core. SystemC not only fits the purpose of system modelling at the Electronic System Level (ESL), but it is enhanced by extensions such as Transaction Level Modelling (TLM), which abstracts communication, and SystemC-AMS, which includes solvers and Models of Computation (MoCs) to model analogue behaviour.

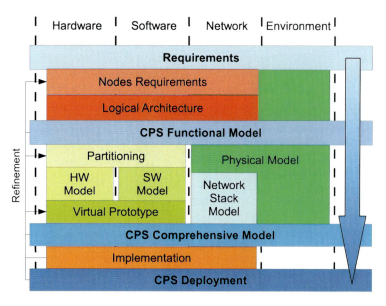

Fig. 8.2 Methodology diagram

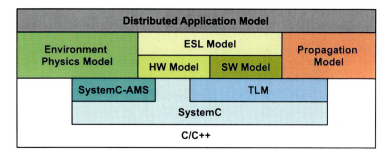

Fig. 8.3 Architecture of the CPS comprehensive model

Furthermore, SystemC is C/C++ based and therefore can be extended with more specific implementations. The architecture of the whole CPS simulation environment is depicted in Fig. 8.3.

Apart from obtaining a comprehensive multi-*domain* model of the system, it is crucial to achieve a multi-*level* model, which adapts the level of abstraction of every model component in order to achieve the best performance possible for the accuracy requirements.

The usage of more or less detailed models depends on what the specific use case demands. E.g. in the case of battery powered nodes where batteries cannot be replaced, such as in a Tire Pressure Monitoring System (TPMS), a very detailed model of the microcontroller might be necessary in order to accurately estimate its power consumption [18].

On contrast, in the use case presented here, although low power consumption is desired, it is not critical since the sensor/actor nodes controlling the appliances are connected to the power grid (see [14]). Moreover, since the wireless network lag usually dominates computation times of the processor, there is no other motivation to use a cycle-accurate model and/or an Instruction Set Simulator (ISS), which would be very costly in terms of simulation performance. Nevertheless, the implementation presented here, is extensible in order to use other models, such as an ISS.

8.4.1 Functional Node Model

In the energy management CPS the main components are an Energy Management Unit and the SmartCoDe nodes, which are those responsible of sensing and controlling the appliances. Sometimes, it can be necessary to use additional very simple sensor nodes just to deliver extra sensed values to the SmartCoDe nodes, when measures have to be done separated from the location of the controlling unit, e.g. fridge. This section will focus on the SmartCoDe nodes.

Wireless physical and MAC layers are both implemented in hardware as part of the transceiver. A transceiver model has therefore been implemented with transmission and reception parameters such as transmitting power, data rate or sensitivity, as well as a state machine which defines what state transitions can be triggered.

Concerning the sensors, important parameters to model are those about data granularity and time required for sampling and value settlement.

Three different kinds of sensors are supported by the sensor interface:

- **Periodically triggered sensors** write sensor values into the event queue periodically (the interval is sensor specific)
- **Single shot sensors** write exactly one sensor value to the event queue when activated (after exhibiting a sensor specific delay)
- **Externally triggered sensors** writes one sensor value to the event queue triggered by an event from the physical domain

The interface is therefore designed to support all sensors present in building automation, as well as some specific sensors inside white goods with relevance to their function as virtual energy storages. What is explicitly not abstracted is the mapping formula from physical dimension to sensor value. While the ZigBee standard is taken as guideline, other cases have to be handled by the application itself.

The actuator interface is modelled after a single analogue output, represented by unsigned 8 bit integer, which is sufficient to describe different power states or dimming levels the appliance can be switched to.

8.4.2 Embedded Platform Model

Along with this functional model development, the embedded platform to be used was selected. This platform is the NXP/Jennic JN5148, which is a System-on-Chip which integrates the microcontroller and the transceiver. Although the hardware platform design is not part of our implementation, SystemC and TLM are widely accepted and standardized for hardware/software co-design of embedded systems [10].

Thus, in this model implementation, there is no need for a highly-detailed model of the sensor/actor nodes. The simulation framework models then only the *relevant* parts of the hardware (wireless transceiver, sensor and actuator-interfaces) on a high abstraction level, as well as the software.

Hardware is modelled by refining the functional model presented in previous section, adding the specific behaviour and characteristics of the NXP/Jennic hardware. For instance, the transceiver model is completed with the specific state machine and transmitting power of the Jennic transceiver. Furthermore, Jennic high power modules, which have a longer reach, are also considered. With this data, and the propagation model introduced later in Sect. 8.4.3, the network topology can be optimized and verified.

The software API presented to the application level code, functionally resembles JenOS API, the embedded OS of the NXP/Jennic ZigBee platform chosen for the implementation. This way, and by using embedded C coding style, the application software developed in the modelling framework can be reused in the real platform with very little adaptation.

The energy management application to be simulated is written in an event-driven style. This reflects common practice in embedded system programming. Even though the NXP/Jennic platform *does* support multi-threading, the recommended way of writing applications is to use only a small number of threads, usually just one, for application level code. Other threads handle network communication tasks, network originated remote procedure calls (RPCs), as well as hardware related tasks. True to the TLM paradigm, those other tasks are not modelled in detail (as would be the case in hardware oriented modelling), but merely their latencies are accounted for. The OS model for the SmartCoDe node functional model can therefore forego true multi-tasking. Instead, there exists a single thread for the application level code which is driven by a single unified event queue.

The downside of this approach is that CPU utilisation and especially contention is not accurately represented by the model. It was deemed of low priority for the case at hand, since the selected hardware platform (32 bit, 16 MHz) provides ample reserve in that respect.

As depicted in Fig. 8.4, the unified event queue is fed by several event sources: network events (e.g. incoming packets, changed network variables), sensor events (when new sensor values are available) and timer events for delayed or periodic execution.

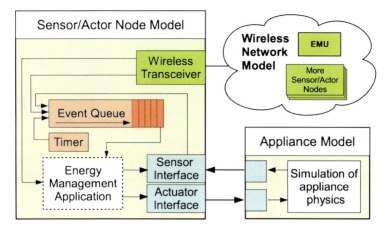

Fig. 8.4 Architecture of the functional sensor/actor node model embedded in the simulation framework

8.4.3 Propagation Model

The simulation includes a radio propagation model realized in SystemC TLM2 [1]. This simulator is capable of modelling noise, collisions, interferences and time-variant effects.

Transaction Level Modeling abstracts communication and separates it from the implementation. Therefore, although it is intended for bus modelling, its approach is suitable for wireless communication too; see [5] for details.

Every node is a module with both an initiator and target socket. The medium is modelled as another module, which acts as a TLM interconnect. In order to be connected, every node has to be registered in the air, which later distributes the data and calculates the attenuation. This registration process binds the TLM sockets.

Transaction Level Modeling is leveraged by abstracting the messages into the transaction data structure, improving simulation performance and providing new means of gathering simulation information in order to evaluate and optimize the system [15].

8.4.4 Network Protocol Stack

The network implements a ZigBee protocol stack. The physical layer and part of the MAC layer are implemented in the transceiver and already introduced in Sect. 8.4.2.

The software protocol stack is also implemented using TLM. Every layer is implemented in an independent software module, with one output and one input interfaces per adjacent layer.

The MAC layer protocol implemented is the unslotted MAC protocol defined in IEEE Standard 802.15.4, which consists basically on a CSMA-CA scheme, with an exponential backoff mechanism between channel assessments. The maximum backoff time is increased exponentially depending on the transmission retries. If the maximum retries are reached and the channel was never clear, the package is discarded.

The model also includes an Ad-hoc On-demand Distance Vector (AODV) routing implementation at the network layer.

MAC and routing protocols have to be implemented and simulated in order to obtain an accurate model of the network behaviour. The number of transmissions carried out by one node is not only governed by the application but also by the number of retries due to the MAC protocol and the number of packages that have to be forwarded because of the routing functionality.

It is crucial for the whole system operation to detect in time a very busy channel where messages can be delivered late or even discarded, with the consequent effect in system performance.

Simulating the routing algorithm is even more important, as an inadequate network topology might cause bottlenecks and overload some nodes with packets to be forwarded. This should be taken into account in application development, but might also affect the hardware production, as some more ZigBee routers can be required in order to solve bottleneck problems.

8.4.5 Environment Interaction

The application to be modelled is the control of appliances providing a thermal service like heating or air-conditioning. The precise appliance class which was modelled captures fridges and freezers, and in the following we will refer to the case of a fridge with a compressor which can be switched on or off, although the approach can be easily extended to more general cases.

It is well known that the temperature profile produced by a fridge can be modelled as a low-pass filter in the time domain (see e.g. [12]). For each power state, the input to the lowpass is the value (in the following referred to as *target temperature*) where the temperature would asymptotically tend to would the appliance stay in that state indefinitely. In the case of a fridge it would be room temperature if it stayed switched off, and a certain temperature usually below $0\,°C$ if it stayed on.

There are formulas to determine the time constant of the lowpass depending on the insulation and other factors, but the goal is to write a controller software which learns the relevant parameters itself without prior knowledge, such that it can be used with *any* kind of fridge/freezer. Therefore, it is more important to have a good *qualitative* temperature model with parameters chosen randomly out of a value-range which produces sensible quantitative temperature profiles.

With SystemC AMS, the thermal model lowpass can be described in a straightforward manner using an embedded Laplace transfer function, e.g.

sca_tdf :: sca_ltf_nd, which describes a transfer function in the numerator-denominator form (for details see [8]), and simulates the temperature in discrete time.

However, two effects which can be observed in real fridge temperature profiles are not yet captured: Usually it takes some time until the switching takes effect in the temperature measured, and the slope of the temperature reverses its direction not abruptly, but gradually. Since the exact parameters of these effects are not only a function of the appliance, but also of the sensor and sensor placement, getting a good qualitative model of this is sufficient. For the case at hand, this has been modelled with a delay counter, which counts down from a given start value after each switch. A turnaround counter, which counts down after the delay counter reaches 0, is used to gradually change the target temperature from the former power state to the new one. Listing 8.1 shows the relevant code of this thermal model.

Listing 8.1 Thermal lowpass model

```
void processing (){
  char input  = in.read();
  if(last_input != input){
    tc = turn_time[input];
    dly = delay[input];
  }
  double t = target_temp[input]
                * tc/turn_time[input]
              + target_temp[1-input]
                * (turn_time[input] - tc)
                 /turn_time[input];
  out.write(ltf_nd( num, den, t, h0 ));

  last_input = input;
  if(dly > 0) dly--;
  else if(tc > 0) tc--;
}
```

8.5 Simulating the Energy Management Application

In a previous paper [6], a partially decentralized energy management approach was described where abstract cost functions are sent to the wireless sensor/actor nodes by an Energy Management Unit (EMU) to steer the power consumption. The nodes then try to control their appliance such that they consume less power at high energy cost intervals and vice versa, and compute a forecast or even a plan of the future power consumption which then is sent back to the EMU for consideration (see Fig. 8.5). Fridges, in that respect, can act as *virtual storages* by cooling down more in low-cost times such that they can switch off in high-cost times.

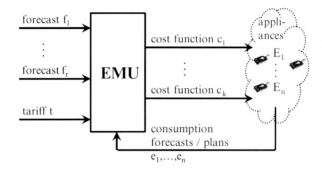

Fig. 8.5 Partially decentralized energy management (Taken from [6])

The cost functions issued by the EMU can be based on various data; apart from the current tariff it might for example include forecasts on availability of solar- and wind-energy. In fact, the latter was a main motivation for the project where this research has been performed. But the cost functions can also be purely abstract just to reach a certain optimization goal. For example, the EMU can send *periodic* cost functions with the same frequency but *phase-shifted* by a certain degree to different appliances for *load balancing* between them; i.e. the EMU tries to achieve that the appliances are switched on at different times such that the overall power consumption is more constant.

Naturally, the problem of writing the software for such an application has two aspects: A global and a local one. Locally the goal is to react to the cost function as good as possible, while at the same time trying to maintain the service of the appliance, e.g. keeping the fridge temperature in between certain temperature bounds. Even if this goal is met satisfactory, this does not ensure that the global goal (e.g. load balancing) is met.

Figure 8.6 shows an overview of the *local* energy management algorithm used after an initial learning phase where usual bang-bang control is used and the normal duty cycle of the fridge is determined in order to parameterize a PI controller. The PI controller then computes an initial schedule (p_{OFF}, p_{ON}) based on the difference (e_u, e_l) of the max/min temperatures (t_{max}, t_{min}) of the last off/on cycle to the temperature bounds (b_u, b_l). This initial schedule is then tweaked according to the cost function. In the case of a periodic cost function, the most straightforward approach for the cost tweak is to search for the nearest cost-minimum. To make sure that the temperature does not go too far out-of-bounds, a temperature forecast algorithm is used which fits a discrete time thermal model to the observed temperature measurements so far.

This algorithm was partly developed in an earlier SystemC model, and then ported to the Jennic/NXP platform and refined further; in fact a real fridge was controlled in lab conditions (A trace of such an experiment can be seen in Fig. 8.7 to the bottom right). To refine the global energy management (a currently ongoing process), the code then was ported back to the SystemC model and used to simulate a group of fridges.

8 Model Based Design of Distributed Embedded Cyber Physical Systems

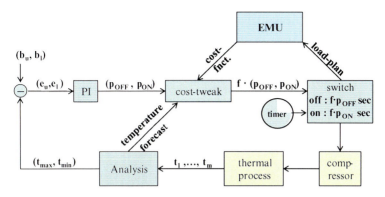

Fig. 8.6 cost-function dependent PI-based fridge controller

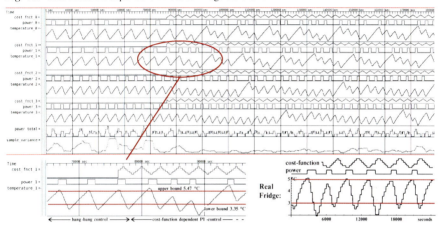

Fig. 8.7 Simulation screenshot, together with an example from a real fridge

Figure 8.7 shows the simulation trace of four fridges being controlled using four periodic cost functions shifted by $\pi/2$. Although there is no feedback to the EMU yet, the combined power consumption is less volatile than in the usual bang-bang case, as can be seen by observing the sample variance. The ultimate goal is to model a given power consumption curve in that way. The simulated time shown is about 2 days, which was simulated in less than half a minute on a normal personal computer.

The use case implementation presented here was mainly used to develop and test an energy management application for energy demand control. In this context, the most important output of the simulation was testing different approaches and algorithms.

The simulation environment permitted evaluating all aspects of the different approaches together, such as network bottlenecks and complexity of the algorithms to be implemented in end nodes. Furthermore, comparisons about what approach performed better in terms of energy management could be established.

8.6 Conclusion and Future Work

This paper presented a model-based design methodology for distributed embedded systems along with a SystemC based simulation framework currently used to develop an advanced energy management application. Despite encompassing several domains and abstraction levels, from high-level processor modelling via medium level wireless network modelling to rather detailed physical temperature models, all could be fit into the SystemC modelling paradigm using the TLM and AMS extensions.

Although modelling every domain and every aspect at every level is still not possible, the framework presented here is flexible enough to enable modelling the domains, aspects and levels of interest, all within the same simulation environment. The only constraints are the simulation performance and the resources required to develop the models.

Both the methodology and the comprehensive simulation proposed have been successfully put into practice in a CPS use case of energy management wireless sensor network, which adapts energy demand of building appliances to both renewable energy supply and environmental conditions.

Future work will include using the framework to further develop the global energy management application, as well as introducing other appliance classes and applications like washing machines or electric car charging, which is currently a hot topic.

Acknowledgements The work presented in this paper has been carried out in the Smart-CoDe project, co-funded by the European Commission within the 7th Framework Programme (FP7/2007–2013) under grant agreement no 247473.

References

1. Accellera Systems Initiative: TLM2.0. http://www.accellera.org (2012, 2013)
2. Accellera Systems Initiative, SystemC AMS working group: SystemC AMS. http://www.systemc-ams.org (2012, 2013)
3. Al-Hammouri, A.T.: A comprehensive co-simulation platform for cyber-physical systems. Comput. Commun. **36**, 8–19 (2012)
4. Bailey, B., Martin, G., Piziali, A.: ESL Design and Verification. Morgan Kaufmann, (2007)
5. Damm, M., Moreno, J., Haase, J., Grimm, C.: Using transaction level modeling techniques for wireless sensor network simulation. In: Proceedings of the Conference on Design, Automation and Test in Europe, Dresden, 2010, pp. 1047–1052
6. Damm, M., Mahlknecht, S., Grimm, C., Bertenyi, T., Young, T., Wysoudil, C.: A partially decentralised forecast-based demand-side-management approach. In: 2nd IEEE PES International Conference and Exhibition on Innovative Smart Grid Technologies (ISGT Europe), Manchester, Dec 2011, pp. 1–7

7. Du, W., Mieyeville, F., Navarro, D., Connor, I.: IDEA1: a validated systemc-based system-level design and simulation environment for wireless sensor networks. EURASIP J. Wirel. Commun. Netw. **2011**, 1–20 (2011). doi:10.1186/1687-1499-2011-143
8. Einwich, K., Vachoux, A., Grimm, C., Barnasconi, M.: SystemC AMS extensions user's guide. Accellera Systems Initiative, SystemC AMS working group (2008)
9. Gajski, D.D., Kuhn, R.H.: New VLSI tools. Computer **16**, 11–14 (1983)
10. IEEE Std 1666–2011 IEEE Standard SystemC Language Reference Manual, (Revision of IEEE Std 1666–2005) pp. 1–638 (2012) doi=10.1109/IEEESTD.2012.6134619
11. Jensen, J.C., Chang, D.H., Lee, E.A.: A model-based design methodology for cyber-physical systems. In: 7th International Wireless Communications and Mobile Computing Conference (IWCMC), Istanbul, 2011, pp. 1666–1671
12. Kupzog, F.: Frequency-responsive load management in electric power grids. Ph.D. thesis, Vienna University of Technology (2008)
13. Lin, J., Sedigh, S., Miller, A.: Towards integrated simulation of cyber-physical systems: a case study on intelligent water distribution. In: Eighth IEEE International Conference on Dependable, Autonomic and Secure Computing, Chengdu, Dec 2009, pp. 690–695
14. Lukasch, F.: Cost efficient mains powered supply concepts for wireless sensor nodes. In: IEEE International Symposium on Circuits and Systems (ISCAS), Rio de Janeiro, May 2011, pp. 502–505
15. Moreno, J., Wenninger, J., Haase, J., Grimm, C.: Energy profiling technique for network-level energy optimization. In: AFRICON, 2011, Victoria Falls, pp. 1–6. IEEE (2011)
16. Ptolemy II: UC Berkeley EECS Depatment, Ptolemy II Project, UC Berkeley http://ptolemy.eecs.berkeley.edu/ptolemyII/ Accessed 2012
17. Tabuada, P.: Cyber-physical systems: position paper. In: NSF Workshop on Cyber-Physical Systems, Austin (2006)
18. Wenninger, J., Moreno, J., Haase, J., Grimm, C.: Designing low-power wireless sensor networks. In: Forum on Specification and Design Languages (FDL), Oldenburg, 2011, pp. 1–6. IEEE (2011)

Chapter 9
Model-Driven Methodology for the Development of Multi-level Executable Environments

Fernando Herrera, Pablo Penil, Hector Posadas, and Eugenio Villar

Abstract Electronic system-level (ESL) methodologies have enabled the development of fast executable system performance models by relying on standard languages such as SystemC. Recent system-level dynamic, that is, simulation-based performance estimation techniques have enabled faster assessment of the design alternatives, and thus the design space exploration (DSE) of complex embedded systems. In this context, the development of system environment models able to reflect common and feasible use cases is crucial for achieving efficient and valid solutions at early design stages. However, such environment modelling can be as or more complex and costly than the system model development itself. The adoption of model-driven development (MDD), component-based design (CBD) and abstraction, can improve the productivity of the environment specification as it does for system specification. In this chapter, a multi-level model-driven methodology for the specification of executable environments is presented. The methodology supports the capture of the environment use cases by relying on the UML standard language and on standard profiles, i.e. MARTE and UTP, and uses UML components for a clean separation of system and environment, and of environment actors. Moreover, a SystemC executable counterpart is automatically generated from the UML-based environment model, coupling the documental and performance analysis levels. The approach is able to capture the communication protocol between system and environment, and also the environment functionality, which can embed either an abstract stimuli generation model, or actual functionality of I/O devices. Thus, different abstraction levels are supported in the functional modeling of the environment.

F. Herrera (✉) • P. Penil • H. Posadas • E. Villar
University of Cantabria, ETSIIT, Santander, Spain
e-mail: fherrera@teisa.unican.es; pablop@teisa.unican.es; posadash@teisa.unican.es; villar@teisa.unican.es

J. Haase (ed.), *Models, Methods, and Tools for Complex Chip Design*, Lecture Notes in Electrical Engineering 265, DOI 10.1007/978-3-319-01418-0_9,
© Springer International Publishing Switzerland 2014

9.1 Introduction

Integration capabilities have undergone a continuous growth (from 10^7 to 10^9 transistors) in the last decade, and further integration capabilities are envisaged [4, 14]. These integration capabilities have led to the possibility of producing more complex embedded systems. However, at the same time, this has involved a major challenge to overcome the gap between design productivity and integration capability. One of the main strategies adopted to overcome the design gap is the development of electronic system-level (ESL) design methodologies [18], where the key initial activity is system specification. Model-driven development (MDD) methodologies enable concepts for making specifications simpler and more understandable, which are major requirements for tackling the design challenge [17]. The usage of standard languages such as UML [23] provides understandability and portability of specifications.

After specification, the next task in an ESL design methodology is design space exploration (DSE) [3, 12, 18]. This activity is crucial for an early assessment of the optimal design decision, since about 90 % of the overall costs are determined in the first stages of the design [12].

Due to the high complexity of the systems and to the huge number of design alternatives, new estimation techniques, such as [8] and native simulation [7, 27], have been proposed for the assessment of the performance of each feasible design alternative. These techniques are dynamic, that is, simulation-based and provide simulation speed-ups of two orders of magnitude with regard to instruction set simulators [27]. Dynamic techniques require the definition of a stimuli environment. Performance results greatly depend on this stimulation, and thus on the design decisions resulting from this assessment. This makes dynamic performance estimation techniques suitable for customized and average optimizations of the systems, which is interesting in many application domains, e.g. wide consumer market, where efficient implementations and cost reduction are crucial. However, these techniques require to enable and facilitate the specification of the system environment as a set of use cases which comprise common cases and corner cases, to let the user dimension the system and assign a given quality of service and guarantees on constraint fulfillment when the system works under both "normal" conditions and under expected worst-case conditions.

However, the development of an executable stimuli environment which can be reused and properly linked to the DSE design flow could be easily more costly than the system specification and the extraction of its executable performance model. Because of this, the specification of the stimuli environment should also support design concepts which have been shown useful for system modeling and design. Adopting an MDD approach for the environment model enables abstraction, and other benefits, such as the application of code generation toolsets for the automated extraction of executable counterparts. This way the whole modelling task, and not only the system specification is covered.

In this paper, a methodology for the abstract modeling and automatic generation of an executable counterpart of the multilevel environment of an embedded system is presented. Specifically, in this methodology, the verification environment is modeled by using UML [23], MARTE [21] and the UML Testing Profile (UTP) [22]. After this stimuli environment model has been developed, a code generator enables the automatic production of an executable SystemC code which reflects all the information captured in the UML model of the stimuli environment. This SystemC model can be easily built up with verification functionality.

The methodology enables the modeling of the environment actors and the specific sequences of interface function calls between the system and the environment actors. This enables to exercise the system, in such a way that depending how the environment actors couple system interfaces, the concurrency of the system application can be more or less exploited, which, in general, impacts on the decision of the optimal mapping to the system platform. It also serves to validate the concurrency structure of the system. These modeling aspects, and the generation of the SystemC executable model were introduced in a previous work [11]. This paper presents the overall methodology, which has been enriched to support additional features. Specifically, the methodology supports now a simplified description of the environment, through implicit sequential diagrams. Moreover, a tool-independent link between the environment model and the files containing the functionality of the environment has been enabled. Finally, the methodology provides now means to describe the environment behavior at two abstraction levels, a first one by capturing an abstract, target independent, description of the environment behavior; and a second, more detailed level, where the system code, typically legacy code, or I/O peripheral driver, is considered part of the environment. This is interesting, for instance, when the system component is use as an input for an automatic synthesis process [28].

The structure of the chapter is the following one. In Sect. 9.2, related and previous work will be presented. Section 9.3 presents the methodology for modeling the environment. Section 9.4 introduces the tooling supporting the environment modeling methodology. Section 9.5 explains how the SystemC model is simulated together with the system performance model. Finally, Sect. 9.6 explains how the methodology has been validated. Sections 9.7 and 9.8 ends with the main conclusions and future work, respectively.

9.2 Related and Previous Work

Several UML-based methodologies for the modeling of an embedded system have been proposed. Intuitively, the modeling of the system environment can be tackled by directly applying the system modeling methodology. However, although maintaining some homogeneity in the modeling methodology of both environment and system can be convenient, environment and system modeling have different constraints and needs. Moreover, certain distinction and asymmetry is required, for instance, to let the implementation framework knows what to synthesize or compile.

At MDD level, this has motivated the development of the UTP standard [22]. Despite the relatively long availability of UTP, only a few approaches have tried to provide support for UTP [15]. In [15] UML and UTP are used for deploying Model-Based Testing in Resource-Constrained-Real-Time Embedded Systems (RTES-RC). This paper aims to close this gap and discusses a concise set of UTP artifacts in the context of model-based testing for RC-RTES. A detailed discussion on the test artifact generation algorithm is presented, demonstrating the applicability of the approach in a real-life RC-RTES example. In [16] the integration of executable uses cases as a supplement to MDD is proposed. It is seen as a model-based approach to requirements engineering. Specifically, a coloured petri-net model is used to express user requirements.

SystemC [13] enables the building of executable, platform agnostic validation environments. SystemC has been widely used for system-level and reusable test bench development, and it has already enabled the development of advanced features for supporting verification and debugging. SystemC has been targeted from several model-based methodologies, focused on the description of a system for the development of executable performance models. Related to verification, in [2] formally sound B models are used to verify model refinement, and translated into SystemC.

The cooperation of fast performance estimation techniques with SystemC has enabled fast simulation of a complex embedded system including SW and custom HW parts. In [19], SW parts are simulated with a virtualization environment called Simics, while SystemC was used for modeling custom HW devices. In [19] the SystemC kernel is made a slave system of the Simics kernel, and an efficient technique for check pointing of the SystemC custom HW was presented. In this approach SystemC is used to model HW devices as an integral part of the system model

The recent merging of the Open SystemC Initiative (OSCI) with Accelera [1] makes targeting SystemC even more interesting, once the proposed modeling environment can benefit from cooperation with other verification approaches, such as the Universal Verification Methodology (UVM) [30].

Although the methodology proposed here does not preclude its extension for supporting a verification methodology, the main motivation was to enable a UML-based, abstract and flexible modeling of the stimuli environment and the automated generation of a standard and executable counterpart. Much of the features of the methodology, presented in [11], were necessary to complete the UML/MARTE COMPLEX modeling and virtual system generation framework [5]. In such a framework, a UML/MARTE-based methodology [9] enables the development of an embedded system model, including the main features in terms of impact on performance. This model can be captured with Papyrus [24], a tool for capturing UML models, which is fully integrated in Eclipse [6]. A related tool, which includes model validators, model-to-text generators [10], and the SCoPE native-based simulation infrastructure [26], enables the automated generation of the performance model. Moreover, the additional features shown in this chapter, e.g. the integration of target-dependent application code as part of the environment, have been applied to a MDD framework enabling automatic software synthesis for many-core framework.

9.3 Environment Modelling Methodology

The proposed environment modeling methodology enables a UML-based modeling of the environment, which can be smoothly integrated in a component-based methodology [29]. Specifically, it is integrated in methodologies such as COMPLEX [5] and PHARAON [25], where the whole system is enclosed in a UML component, and where different views, in the shape of UML packages are used to capture the system model. The proposed specification of the environment, as shown in the following sections, cleanly separates the system information from the model containing environment actors, their functionality and their interconnection with the system model.

9.3.1 Environment Structure and Connection to the System

The user can develop the model of the environment at the same abstraction level, clearly separating the system from the verification environment. Specifically, the user will enclose all UML modeling elements within a specific view of the model: the verification view. Figure 9.1 shows an example with the hierarchy of UML elements used for the modeling of the environment proposed, and which can be taken as a reference for the following discussion. The verification view is actually a UML package, typed with the «VerificationView» stereotype, which contains the model elements which describe the verification environment facilitating a tool-independent separation of system and verification elements. The verification view declares the whole set of actors which compose the environment as a set of UML

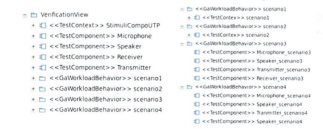

Fig. 9.1 Several scenarios are supported

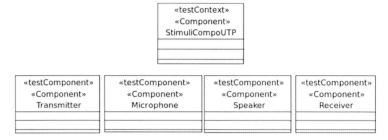

Fig. 9.2 Declaration of environment and test components in the VerificationView

components with the UTP «TestComponent» stereotype applied (Fig. 9.2). An additional UML component, with the UTP «TestContext» stereotype is used for the declaration of a verification environment (StimuliCompoUTP in Fig. 9.2). The internal structure of this component, depicted in the composite diagram in Fig. 9.3, reflects the interconnection structure of the system and environment component instances. Environment component instances are captured as UML properties typed as «TestComponent» components, and the system component instance. The system component is captured as a UML property typed as the UML component reflecting the system, and which exports I/O functional interfaces. Notice that this system component is not in the verification view, but in a view related to the specification of the system description. This means a dependency, so the verification view depends on the system views. In addition, the referred system component must be specified by the UTP «SUT» (System Under Test) stereotype. Through this scheme a clear separation is established between the system element and the environment elements.

The composite diagram in Fig. 9.3 also shows the port to port connection. After this interconnection, the environment components that provide the services required by the system are stated. Similarly, services provided by the system can be invoked from the environment modules.

9.3.1.1 Modelling the Behaviour of the Environment: One Scenario

As well as the interconnection between the environment elements and the system, the proposed methodology supports the specification of the behavior of the environment. First, the methodology adds a main concept, the scenario. A scenario models the activity of the different environment components, and their interaction with the system for a given use case. Several scenarios are possible (see Sect. 3.4). It means that, while one scenario can involve activity in all the environment components, each with a specific behavior, a different scenario can model activity only in some environment actors, with a different behavior. Each scenario can be described by:

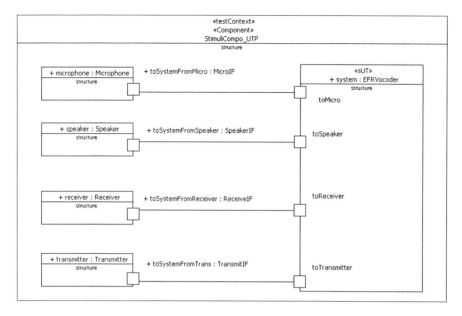

Fig. 9.3 Environment structure for an EFR vocoder

- Interactions, each interaction between an environment component and the system is a totally ordered sequence of service calls, which can be synchronous or asynchronous.
- References to file sets, where the specific functionality is allocated.

A scenario is captured as a UML package stereotyped with the MARTE stereotype «GaWorkloadBehavior». A scenario package has to be a child element of the VerificationView package. The methodology does not enforce the use of both type of functionality. Indeed, the UML/MARTE COMPLEX methodology uses only interaction diagrams, such a later code generation phase produce templates with empty functionality. Therefore, filling the functionality is left to the user, as a manual task. Instead, in the context of the UML/MARTE PHARAON methodology, file references are included, while interaction diagrams are omitted. In practical terms, it involves two different environment modelling styles. However, both of them fit to the more general scheme presented here.

Interaction Modeling

Interaction modeling relies on UML interactions, where UML lifelines can represent either, the system or an environment component. As a prerequisite, the GaWorkloadBehavior package (scenario package) must comprise a UML component with the UTP «TestContext» stereotype. Then, this new

Fig. 9.4 Generalization for referencing environment components in the modeling of each scenario

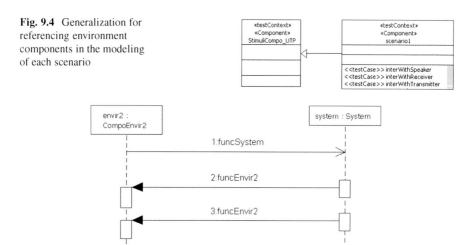

Fig. 9.5 Sequence diagram for a port to port interaction

«TestContext» component is generalized by the «TestContext» component where the environment structure is specified (Fig. 9.3). This «TestContext» components relation is modelled by an UML generalization (Fig. 9.4). This way, the component instances reflecting environment components can be accessed and later on associated to UML lifelines, used for specifying scenario interactions.

A scenario comprises the specification of all the interactions over time between the system and the environment components. They are described by means of one or more UML interactions (as child elements of the «TestComponent» component) which also have the UTP «TestCase» stereotype applied. A scenario description is complete when all the interactions cover all environment components and their ports. However, this is not a required condition since a scenario can represent a use case which might not require an interaction with all system ports. UML interactions are graphically captured by means of sequence diagrams. Figure 9.5 shows a sequence diagram capturing the interaction between an environment component and the system component. A lifeline references the instance of the system component, while the other lifeline references an instance of one environment component. Making these references is feasible thanks to the specialization shown in Fig. 9.4. As well as the lifelines, the interaction contains the set of UML messages exchanged between the system and the environment component. These messages represent function calls, as services provided either by the system to the environment or viceversa.

The different environment components are communicated with the system by using interfaces and specified by the MARTE stereotype «ClientServerSpecification» and they contain the functions used for the component interconnections. The interfaces are included in the model view «FunctionalView». Depending on the goals of the designer, these interfaces can represent auxiliary interfaces used for defining functions for validating the concurrency and behavior of the system in different use cases. After the system validation using the stimuli specifications, the designer can develop the environment interfaces for physical implementation in order to access to the environment actors which represent peripherals. In this case, these interfaces are the implementation mechanisms for accessing these peripherals according to predefined functional and non functional requirements.

The sense of a UML message is captured through its "from" and "to" attributes (in the diagram, the "to" attribute corresponds to the tip of the arrow). The sense states whether the system calls a function provided by the environment ("from=system") or, on the contrary the environment requires a service provided by the system ("to=system").

Two different types of UML messages are used: synchronous messages and asynchronous messages. They enable the specification of synchronous and asynchronous services. Synchronous services require the return of the function call, e.g. because the client expects some output data from the service call. However, it might be interesting to specify service calls which just provide input data for triggering the service, and which immediately return and let either the system or the environment component go on executing. A UML synchronous message is represented by a solid arrow head (as in the two messages from the system to the environment component in Fig. 9.5). UML asynchronous messages are represented by open arrow heads (as in the message from the environment component to the system in Fig. 9.5).

The name of the message identifies which function is called. This is required because an interface can comprise several service functions. In the example of Fig. 9.5, the environment component first calls the funcSystem service provided by system component. Next, the system calls the "funcEnvir2" service twice provided by the environment component. These functions must be part of the interface accessed through any of the environment component ports. And, as explained, those interfaces could present more functions, e.g. "funcSystem2" or "funcEnvir1".

Although the sequence diagram graphically reflects a total order in the exchange of messages, this information is not contained in the UML interaction. Diagrams provide a graphical representation, but not all that graphical information is contained in the UML model. It is required to add this information in a way it can be preserved in the .uml file read, so available to the toolset around the model, e.g. model validation or code generation frameworks. For it, in the proposed methodology, a unique order identifier ("i:") prefixes the message name, which is part of the UML model. In this way, a total order in the exchange of messages can be specified at local level. Local level means the interaction between the system and a single environment component, which requires two lifelines, as in Fig. 9.5.

The user can specify all the interactions in a compact way. In fact, the messages of a sequence diagram can refer to functionalities of different ports of the environment component, and thus of the system. Moreover, one UML interaction can be used for specifying the interaction of the system with more than one environment component. In Fig. 9.6 a sequential diagram shows the communication between the system and two environment components. In principle, in a diagram like this, the sequence of messages exchanged between the system and one environment

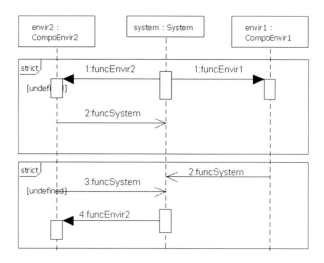

Fig. 9.6 A sequence diagram stating synchronization conditions between the systems and two environment components

component is not related to the sequence of messages exchanged between the system and another environment component. That is, in the example of Fig. 9.6, if the reader forgets by now the strict labeled boxes, there would be in principle no order relationships between the messages exchanged between system/envir2 lifelines, and the messages exchanged between system/envir1 lifelines. That is, the i-th message of system-envir2 communication might happen before, at the same time, or after the j-th message of system-envir1 communication.

However, use cases may actually require the modeling of these types of constraints, because the environment itself can also present dependencies, e.g. among environment components, and thus provoke dependencies between system interfaces which do not have its origin in the system itself. In the proposed methodology, the user can specify order relationships among messages exchanged by different environment components with the system. This is done by using UML CombinedFragments, shown as boxes in Fig. 9.6. Specifically, in the Fig. 9.6 example, a strict combined fragment is used. The strict combined fragment groups the execution of the set of messages it covers, so that all covered messages have to be executed before or after the remaining messages. That is, it defines an atomic region of messages exchange. Taking the previous discussion into account, the use of combined fragments adds a higher ordering level to the specification of the environment, in the sense that all the messages encapsulated in the same combined fragment are associated with a single and higher order implicit ordering index. Moreover, it also adds a global ordering since it covers the interaction of the system with more than one environment component.

For instance, the two combined fragments in Fig. 9.6 state that "1:funcEnvir2", envir2 call to "2:funcSystem", and "1:funcEnvir1" will have an associated higher order k-th index; and will have to be executed before or after "3:funcSystem",

"4:funcEnvir1", and envir 1 call to "2:funcSystem" messages, with a m-th high order index. The total order of each local system-environment component interaction, imposes a total order in the execution of the combined fragments ($m < k$). In other words, the bottom combined fragment in Fig. 9.6 has to happen after the top combined fragment. Local and global ordering has to be coherent, thus two environment component lifelines cannot impose an order conflict on two combined fragments (e.g., $m < k$, $m > k$). As a result, the methodology enables the specification of a partial order of messages exchange. A partial order of messages is specified where there can be order relationships among the messages sequences exchanged by the system with different components. Specifically, in the Fig. 9.6 diagram, "1:funcEnvir2" and "2:funcSystem" messages, reflecting function calls among envir2 environment component and the system, will take place before the "2:funcsystem" call done by envir1 environment component. Similarly, the diagram in Fig. 9.6 specifies that "3:funcSystem" and "4:funcEnvir1" will take place after "1:funcEnvir1".

The methodology also supports another two combined fragments. The "loop" combined fragment (loop) is used for specifying repetitive subsequences of message exchange. The "parallel" combined fragment (par) is used to model that certain groups of services either provided or required by the same environment component can be executed in parallel.

The features presented up to here enable the specification of a partial order of service calls in the environment. Formally speaking, this is the most abstract way to specify time constraints in the environment model. Furthermore, the proposed methodology enables the association of physical time information with the environment model. Specifically, the initiation of each service call can be placed in a specific physical time stamp. In order to specify it, the MARTE «TimedProcessing» stereotype is used. This stereotype is applied to the UML message which reflects the service call placed in physical time. The stereotype provides the attribute "start", which denotes a UML Time Event, which in turn, is placed in physical time through a UML Time Expression.

Implicit Interactions

The methodology admits the use of implicit interactions. It is a practical feature which saves time and complexity in the modeling of common environment models in a specific domain. Specifically, it means that an environment component will have a default interaction scheme associated, if no specific UML interaction has been captured and associated with it. A methodology can define this implicit interaction. For instance, in a domain space oriented methodology, hard real-time analysis methodologies will typically assume a reactive environment, and an active system which does not block because of waiting for environment services, whose response might be non-predictable and/or unbounded. It is typically modeled through an interaction scheme with an infinite loop enclosing an incoming asynchronous UML message. I.e. requested by the system. Therefore, this interaction scheme is a good implicit interaction candidate for space domain oriented applications.

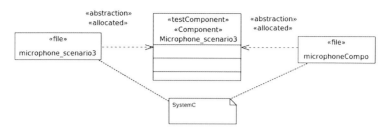

Fig. 9.7 File association with a TestComponent

Association of File Sets

The proposed methodology supports the capture within the environment model of a file set related to an environment component. All the interface functions present in the interactions of the environment component could find its implementation in the file set or not. In the former case, this feature enables a fully automated integration of source code, e.g. previously implemented test benches in SystemC or C/C++, with the rest of the environment model information. The association is also applicable to the case of relying on implicit interactions. In such a case, the functions called can be inferred from the associations or be explicitly captured as environment component operators. The scenario when this happens is explained later on. As mentioned, it might also happen that not all the functions associated with the environment component are found in the associated file artifacts. In such a case, code generators produce the interface function declarations and the implementation templates.

Figure 9.7 shows how file set association is captured in the proposed methodology. Let assume that environment functionality is available in a set of source files. The file set is modeled as a UML artifact typed by the UML standard stereotype «File». These file artifacts are included in the «FunctionalView» package. These files represent previously created test-benches and, thus, can be reused in different designs or the user code of the component functionality.

When the environment component functionality is specified by files the corresponding GaWorkloadBehavior package should contain additional TestComponent components. These TestComponent components are generalized from the TestComponent defined in VerificationView package (Fig. 9.8). These new TestComponent have associated the different files where the functionality of the scenario is implemented. This file-TestComponent association is modeled by a UML abstraction specified by the MARTE stereotype «Allocated» (Fig. 9.7).

Fig. 9.8 TestComponent generalization

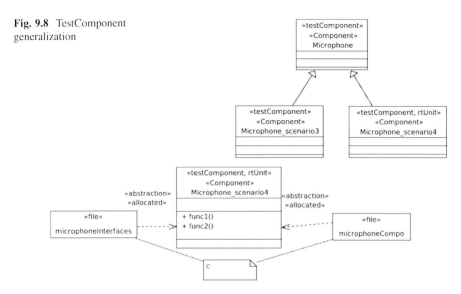

Fig. 9.9 TestComponents as application components

9.3.2 Levels of Abstraction in the Specification of Environment Behaviour

The proposed environment modeling methodology supports de distinction at the modeling level between an abstract (target independent) and a target dependent description of the environment behaviour.

Figure 9.7 exemplifies the modeling of the former case. Then, the model reflects that the source code linked to the environment model reflects a functional model of the environment and which can be therefore used for creating an executable counterpart of the environment model. This is independent from the language. However, a language such as SystemC is typical of this case, since SystemC is a language suitable for implementation agnostic models. Moreover, the proposed methodology supports a system-dependent, and more specifically target dependent, description of the environment component. The case is show in Fig. 9.9. The idea is that methodologies, require the consideration of certain application components as environment components. For instance, in PHARAON methodology, there are application or platform software components which reflect the software layer for accessing I/O peripherals. However, it is not interesting to consider them as part of the system, e.g., for code synthesis effects. However, they can contain C or C++ code reflecting the functionality of the device driver which facilitates and makes realistic the development of the environment, or that can even enable hardware in the loop methodologies (enabling the integration of the peripheral hardware, e.g. a camera, together with its driver, as part of the environment, and the rest of the system under design).

A distinctive aspect of this case is that the source code ("C" code in Fig. 9.9 example) is likely target dependent code. E.g., a high-level device driver code used as part of the test bench will likely make calls to an operative system instance, which in turn is part of the system under design, and thus such operative system instance is captured in the system specification. Therefore, this model needs to break the pure separation of scopes between the environment and the system. As this is not the general rule in a MDD methodology, it is convenient to mark the environment components which such access to the system model internals. It is shown in Fig. 9.9, where the environment component is typed by the «RtUnit» MARTE stereotype, as well as by the UTP «TestComponent» stereotype.

9.3.3 Modeling Several Scenarios

The proposed methodology supports the modeling of several GaWorkloadBehavior packages. This enables a set of different stimuli associated with different use cases to be captured in a single model. Then, a DSE exploration can be done for each use case, and the design can be tailored for a set of use cases. A a wide set of scenarios can be also used for validating a single design for different use cases. In addition, the methodology provides different GaWorkloadBehavior packages where the file association is used. This fact enables the verification of the system with test-bench files (SystemC or C/C++) and so the verification of the peripheral interfaces for the final implementation. In this way, both modeling mechanisms can cohabit in the same VerificationView package which enables the definition of the different design stages in the same model. In order to specify several scenarios, the user only needs to specify a new GaWorkloadBehavior package, in turn containing a TestContext component. This TestContext component again generalizes the TestContext component and owns as many interactions as the user requires for describing the new scenario.

9.4 Toolset

The environment modeling methodology presented is partially supported by a toolset which relies on Eclipse and Papyrus. Specifically, two code generation tools have been separately implemented up to now. The first generator produces a SystemC counterpart from the ULM interactions. The second one, implements the generation of the file structure from the model. These generators have been written in the standard Model-to-Text (M2T or MTL) language [20], to improve its portability across different model-to-text transformation engines.

9.4.1 SystemC Generation

The code generation is in charge of producing all the SystemC code reflecting the structure of components and concurrency present in the UML/UTP/MARTE environment model. It also produces the service calls fulfilling the partial order specified in the environment model by means of sequential diagrams. The generator does not produce functional code, whose insertion is left to the user. However, in order to enable the production of an executable environment model from the first moment after the generation, void functions with debugging printouts are produced. This permits a fast initial check of the SystemC code produced and provides a basis for indicating to the user where to insert functional code.

Code generation is actually done in two phases. First a model-to-text transformer translates the UML environment model into a set of macros. A specific front-end of the SCoPE simulation framework provides the SystemC translation for these macros. The SCoPE framework enables the compilation of a dynamic library (instead of a static executable file) for each scenario. In this way, the approach is modular at executable level, in the sense that each scenario of the SystemC environment has its own .so file, separated from the .so files of the system executable model itself. The generator basically maps all verification views to a single SystemC module (thus there is no UML environment component module mapping). The code generator produces at least one SystemC process containing a sequence of channel accesses for each environment component. This sequence fulfils all the order constraints specified through the sequence diagrams, as explained in Sect. 9.3.

9.4.2 File Structure Generation

The other implemented code generator enables the generation of the file structure. In order to characterize an application component, the files whose functionality is implemented, the interfaces required/provided and the component's functions should be specified. With this information, the code generator creates the application files (.c/cpp and .h). These files include the declaration of the functions provided by the application component through its interfaces and the other functions specified in the model as internal component functions. The functions of the application environment component can be associated with a specific file in order to specify that a function has to be included in this file. Otherwise, the code generator produces two additional files, apart from the files specified in the model, one which includes the declaration of the functions of the interfaces provided by the component and another file which includes the internal functions of the component. Then, by using a UML comment, the programming language is annotated (Figs. 9.7 and 9.9).

In addition, the second generator enables the generation of the makefiles required for the compilation of the environment components with the rest of the application in order to be executed in the simulation tool. This feature enables the designer to focus on the functionality implementation and not on the infrastructure required for the simulation tool execution.

9.5 SystemC Simulation with the System Performance Model

The system simulation infrastructure used (SCoPE+) works on top of the SystemC kernel. Time advance, buses and peripherals have been developed using the standard SystemC features. As a result, no kernel synchronizations between the system simulation kernel and the environment kernel are required, which contrasts with other approaches such as. However, direct use of SystemC environments is not possible since both parts rely on different models of computation.

On the one side, the system model uses a client-server component-based communication, based on function calls. Each time a client component requires a service, it calls a function that is implemented and provided by another provider component. Under this perspective, the data transferred among components are the input and output arguments of the functions of component interfaces. Each function called by a client component means sending input arguments to server components and, if output arguments are expected, they are sent from the m server component to the client component. On the other side, the SystemC environment relies on interfaces based on transfers which use SystemC channels. These channels receive packaged data and provide different communication semantics: blocking/non blocking, with/without memory, etc.

Interconnection wrappers are used to adapt each channel of the SystemC environment to each function in the system interface. In these wrappers, communication accesses are divided into two steps: a request step, where the input arguments are sent, and a response step, where output arguments return to the calling task. In the meantime, the calling task is blocked, waiting for the response. This approach models the blocking nature of function calls. Additionally, data transfers are packaged by copying the arguments of each function call in a buffer that is sent through the SystemC channel as a single unit.

9.6 Example

The suitability of this methodology has been demonstrated through the development of environment models for an EFR vocoder example. The interrelation between the environment components and the system has been specified in two different ways, by using the UML interactions and by means of the files. The SystemC code was automatically extracted from the environment and simulated with the executable performance model, automatically extracted from the UML/MARTE model of the system, after requiring only the injection of the functional code.

Figure 9.3 shows the structure of a first environment model developed for simulating a full-duplex transmission operation mode. In this use case, the coder and the decoder functionalities of the vocoder are stimulated independently (and thus potentially at the same time) and they have to exhibit concurrent behavior able to attend to coding and decoding services at the same time. This model is composed of

9 Model-Driven Methodology for the Development of Multi-level... 161

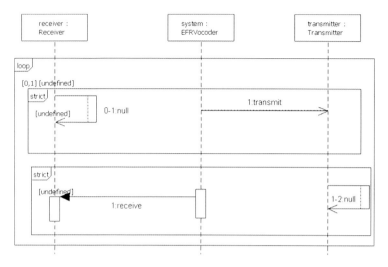

Fig. 9.10 Modeling of order constraint for remote closed loop modeling

four environment components. In this way, independent threads stimulate the coder and decoder within the vocoder, and the collection of coded audio (transmission) also runs independently of the collection of the decoder output (speaker output).

In order to show the usage of more than one scenario, the environment with the structure shown in Fig. 9.3 was extended to support a second scenario, specifying a remote close loop use case. In principle, this could also be done with the structure shown with the addition of some functionality for emulating channel effects. The only additional requirement to cover the specification needs of "scenario2" was the ability to specify that the service for transmitting the i-th coded frame should always be called before the reception of the corresponding i-th received frame, reflecting the same coded frame but corrupted by the channel effect after traversing the remote closed loop. Notice that since different components (and processes) are inferred for transmitter and receiver, and since the receiver can directly read from a file with the coded and corrupted frames, the ordering constraint is required to model the actual causality existing between the i-th coded frame sent and the corresponding received frame after traversing the remote loop.

Figure 9.10 shows the sequence diagram which reflects the interaction of transmitter and receiver with the system. Two strict combined fragments are used to reflect the aforementioned order condition, which will impose an order relationship in the test-bench generation. The combined fragments cover lifelines from both the transmitter and from the receiver environment components. Notice that the null activity of one environment component while the other is attending to a service is explicitly modeled. The sequence diagram in Fig. 9.10 also illustrates the usage of synchronous and asynchronous messages for modeling synchronous and asynchronous service calls. In fact, the EFR Vocoder uses the asynchronous transmit service since the vocoder can go on coding frames after the coded audio frame is delivered to the transmitter. However, the vocoder uses a synchronous call for the services to be called since the decoder cannot work without having received the output parameter of the received function, that is, the received audio frame.

The scenario "scenario3" (Fig. 9.1) is part of a test-bench collection to be used for checking the vocoder system. In this case, it is not necessary to specify the interactions for functionality description; only the capture of the different files in the model is required. The infrastructure for system simulation (makefiles) is automatically obtained for the SCoPE+ simulation.

Finally, the scenario "scenario4" (Fig. 9.1) specifies the target dependent source code (Fig. 9.9) of the different device drivers for the code synthesis process. The makefiles generated included all the information required for compiling the application in the target board (cross compiler, flags...).

9.7 Conclusions

Support for MDD and related tools in the specification of a stimuli environment is necessary for the development of performance models for complex embedded systems. It enables fast model development and efficient design decisions in the DSE phase. This paper describes a methodology for UML/MARTE/UTP modeling of an environment which supports the specification of the main environment actors and their interconnection with the system; the specification of the interaction of environment components with the system as partially ordered sequences of service calls; and the specification of several scenarios for reflecting different use cases. In addition, the methodology enables the capture of the files which implement the functionality of test-benches for system simulation in different scenarios. This avoids modeling the environment-system interactions in order to take advantage of the previously implemented test-benches. Moreover, the methodology enables the specification of peripheral interfaces to be developed for the final system implementation which, by using the simulation, enables the verification of the interface's functionality required for the final system synthesis implementation.

Tools support the generation of the SystemC code and the makefiles infrastructure for execution in the simulation tool.

9.8 Future Work

Some methodological aspects have still not been implemented in a specific tool. Specifically, there is no support for the generation of the function calls sequence which defines the behavior of an application environment component in a specific scenario. This generator would generate the complete, ordered sequence of function calls established between the system and the environment application component. This sequence of call functions would be included in a file. This file is automatically generated. However, it is possible to define where these function calls should be allocated in a specific file that has previously been captured in the model. In order to do so, a UML operation should be specified in the corresponding application environment component. Then, this operation is associated with a file artifact. In this way, the sequence of function calls which composes the communication statements are specified in the body of this function allocated in the file artifact. Finally, a complete framework which integrates all the code generators and enables all the environment specification and simulations is still to be implemented.

Additionally, the environment modeling methodology could be extended. A natural extension of this work consists in the addition of verification capabilities, by using further UTP stereotypes for the specification of assertions, supported by an extension of the validation tool.

Acknowledgements This work has been funded by the European FP7-247999 COMPLEX project, FP7-288307 PHARAON project and by the Spanish MCI TEC2011-28666-C04-02 DREAMS project.

References

1. Accellera: http://www.accellera.org/home/ (2013)
2. Cansell, D., Culat, J.F., Méry, D., Proch, C.: Derivation of SystemC code from abstract system models. In: Proceedings of FDL 2004, Lille, Sept 2004
3. Chang, H., Cooke, L., Hunt, M., Martin, G., McNelly, A.J., Todd, L.: Surviving the SOC Revolution: A Guide to Platform-Based Design. Kluwer, Boston (1999)
4. Chiang, S.Y.: Keynote speech. In: Proceedings of ARM Techcom Conference, Santa Clara, Oct 2011
5. COMPLEX Project: http://complex.offis.de (2013)
6. Eclipse project website: http://www.eclipse.org/ (2012)
7. Gerin, P., Hamayun, M., Petrot, F.: Native MPSoC co-simulation environment for software performance estimation. In: Proceedings of the CODES+ISSS'09, Grenoble, Oct 2009
8. Gligor, M., Fournel, N., Pétrot, F.: Using binary translation in event driven simulation for fast and flexible MPSoC simulation. In: Proceedings of the CODES+ISSS'09, ACM, Grenoble, France (2013)
9. Herrera, F., Peñil, P., Villar, E., Ferrero, F., Valencia, R.: An embedded system modeling methodology for design space exploration. In: Jornadas de Computación Empotrada (JCE), 2012. Alicante, Jornadas Sartenco. Elche, Sept 2012
10. Herrera, F., Posadas, H., Villar, E., Calvo, D.: Enhanced IP-XACT platform descriptions for automatic generation from UML/MARTE of fast performance models for DSE. In: DSD, Izmir, Turkey 2012
11. Herrera, F., Penil, P., Posaads, H., Villar, E.: A model-driven methodology for the development of SystemC executable environments. In: Proceedings of the FDL 12, Viena, Sept 2012
12. Holzer, M.: Design space exploration for the development of embedded systems. Thesis dissertation, Vienna University of Technology, Vienna (Apr 2008)
13. IEEE Std. 1666-2011: IEEE Standard for SystemC® Language Reference Manual. http://standards.ieee.org/getieee/1666/download/1666-2011.pdf (2012)
14. Intel 22 nm Technology: http://www.intel.com/content/www/es/es/silicon-innovations/intel-22nm-technology.html?wapkw=22nm (2013)
15. Iyenghar, P., Pulvermueller, E., Westerkamp, C.: Towards model-based test automation for embedded systems using UML and UTP. In: IEEE 16th Conference on Emerging Technologies And Factory Automation (ETFA), Toulouse, Sept 2011, pp. 1–9

16. Jogesen, J.B.: Executable use cases: a supplement to model-driven development? In: Model-Based Methodologies for Pervasive and Embedded Software, MOMPES, Braga, Portugal 2007
17. Kopetz, H.: The complexity challenge in embedded system design. In: 11th IEEE ISORC, Orlando, May 2008
18. Martin, G., Bailey, B., Piziali, A.: ESL Design and Verification: A Prescription for Electronic System Level Methodology. Systems on Silicon. Morgan Kaufmann Publishers Inc., San Francisco, CA, USA (2007). ISBN: 9780080488837
19. Monton, M., Gladigau, J., Haubelt, C., Teich, J.: Checkpoint and restore for SystemC models. In: Borrione, D. (ed.) Advances in Design Methods from Modeling Languages for Embedded Systems and SoCs. Springer, Dordrecht/New York (2010)
20. OMG: MOF Model to Text Transformation Language (MOFM2T), 1.0. http://www.omg.org/spec/MOFM2T/1.0/ (2008)
21. Object Management Group. UML profile for MARTE: Modeling and Analysis of Real-Time Embedded Systems. Version 1.1. (2011). Available in http://www.omg.org/spec/MARTE/1.1/. Accessed 2013
22. Object Management Group. UML Testing Profile (UTP). Version 1.1. (2012). Available in http://www.omg.org/spec/UTP/1.1/. Accessed 2013
23. OMG Unified Modeling Language: Infrastructure and Superstructure. V2.4.1. www.uml.org (2013)
24. Papyrus: http://www.eclipse.org/modeling/mdt/papyrus/ (2012)
25. PHARAON project web: http://pharaon.di.ens.fr/ (2013)
26. Posadas, H., Real, S., Villar, E.: M3-SCoPE: performance modeling of multi-processor embedded systems for fast design space exploration. In: Silvano, C., Fornaciari, W. Villar, E. (eds.) Multi-objective Design Space Exploration of Multiprocessor SoC Architectures: The MULTICUBE Approach. Springer, New York (2011)
27. Posadas, H., Díaz, A., Villar, E.: Annotation techniques and RTOS modeling for native simulations of heterogeneous embedded systems. In: Tanaka, T. (ed.) Embedded Systems-Theory and Design. Intech, Rijeka (2012)
28. Posadas, H., Penil, P., Nicolás, A., Villar, E.: Automatic synthesis of embedded SW from UML/MARTE models based on memory space definitions. In: Design of Circuits and Integrated Systems (DCIS), Avignon, France 2012
29. Szyperski, C.: Component Software: Beyond Object-Oriented Programming, 2nd edn. Addison-Wesley Professional, London (2002)
30. Universal Verification Methodology (UVM) 1.1 Class Reference: http://www.accellera.org/downloads/standards/uvm (2011)

Chapter 10
GREEN HOME: The Concept and Study of Grid Responsiveness

Slobodanka Tomic, Jan Haase, and Goran Lazendic

Abstract This paper describes the concept for a distributed demand management program based on bi-directional communication involving an energy management system of a Grid Responsive Energy Efficient Networked (GREEN) home and the energy supplier in the Smart Grid. Our work is motivated by expected benefits that an energy supplying company can achieve based on additional high-granular knowledge regarding his customers' consumption habits. Our demand management approach uses in-home energy consumption monitoring and forecasting of future demands, which are in aggregated form available to the supplier as a day-ahead and intra-day forecasts. The supplier uses these forecasts to achieve higher precision of trading, and consequently reduce total energy cost. This trading gain he can share with his customers participating in such a program.

10.1 Introduction

The communication and information technology is a main building block of novel energy management solutions, which promise to make energy consumption in households smarter in terms of eliminating energy waste and reducing the energy bill [1]. In a vision of the next generation smart home, an autonomous decision support system for smart home management is able to associate each of the home appliances, sensors or devices with some functions, such as cooling, heating,

---

S. Tomic (✉) • G. Lazendic
FTW Forschungszentrum Telekommunikation Wien GmbH, Donau-City-Straße 1/3,
A-1220 Vienna, Austria
e-mail: tomic@ftw.at; lazendic@ftw.at

J. Haase
Vienna University of Technology, Institute of Computer Technology, Gußhausstr.
27-29/384, A-1040 Vienna, Austria
e-mail: haase@ict.tuwien.ac.at

J. Haase (ed.), *Models, Methods, and Tools for Complex Chip Design*, Lecture Notes
in Electrical Engineering 265, DOI 10.1007/978-3-319-01418-0_10,
© Springer International Publishing Switzerland 2014

light, entertainment, and with people in the home, their preferences, actions and needs. Based on this model, such system can automatically control some of the home functions using the preferences of users, e.g., regarding the required level of comfort, and optimize the energy use or energy bill by reducing consumption or using cheaper energy [2]. As some of the energy consumption decisions will always be under the direct control of people, who can also override all automatic control decisions, the home management system must be able to re-optimize the system after every user action.

The smart home vision is driven by two trends: on the one hand, a broad variety of intelligent personal or home devices or appliances will be able to simply plug-in into the home communication network, and to communicate with the external world, to send their data or directly receive remote control messages [3]; on the other hand, there is an effort to establish one single device, the Home Gateway as a central intelligence component, coordinating not only the in-home communication, but also acting as a central decision making component for home automation and energy consumption management and optimization [4–10]. For the purpose of energy management in the future smart homes such Home Gateway is envisioned to communicate with the Smart Grid, for example directly or through the Smart Electricity Meter, facilitating a non-discriminative access to the home for different energy suppliers and other Smart Grid service providers.

Indeed, different intelligent energy service providers may be interested in establishing programs for demand management or demand response aiming at enhancing the grid stability through controlled load shedding, or managing the demand through dynamic prices.

In our vision of energy efficient homes, the precise monitoring of home energy consumption conducted within the home is regarded as one central system functionality. We assume that the Home Gateway will keep the track of consumption by reading the Smart Meter's data directly. We also assume that all appliances and other home equipment that consume energy will soon be capable of measuring their own power consumption, detecting their own state, and communicating these to the Home Gateway. Today, similar functionality is available via smart plugs.

The other central functionality of the home management system is to understand how home residents interact with different home devices, and how flexible they are in these interactions. The user-centered optimization of energy consumption requires that users' utility of some equipment as a function of time is put into the relation with the price of energy. The optimization potential grows when it is possible to automatically detect intervals in which devices are switched on but are essentially only wasting energy, e.g., lights are on but no one is at home, and from detecting potential to schedule some devices away from the periods of general peak consumption, into periods with high energy availability. Load schedules optimized only in respect to Smart Grid constraints may potentially not be acceptable from the point of view of the users, as they do not satisfy the constraints that come from their individual flexibility. For example a user may be restricted in accepting a schedule that starts his washing machine because of some rules that apply to all tenants of the apartments in his building.

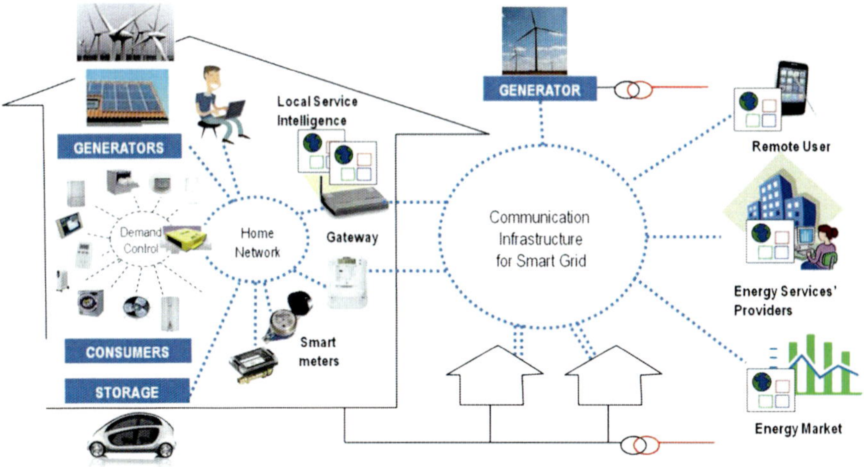

10 GREEN HOME: The Concept and Study of Grid Responsiveness

Fig. 10.1 GREEN Home overall scope

The concept of *grid responsive energy efficient networked home* integrates a number of information and communication technologies to achieve energy efficiency and costs savings based on active customer participation and automation. The overall scope and the general architecture of the GREEN Home System are illustrated in Fig. 10.1.

While in our vision of grid responsive homes the smart home system may support interactions with different Smart Grid actors, such as energy suppliers, distribution system operators, new energy markets or other energy service providers, in this paper our special focus is on functions that support demand management governed by an energy supplier. Today, the energy supplying companies purchase energy for their customers using standardized user profiles [11], as they do not obtain real-time consumption data of their customers. The energy is purchased through long-term procurements and short-term trading. In particular the peak demand is satisfied by means of short term procurements mostly at higher prices. While standard profiles provide aggregated models of different levels of demand during different periods of the day and time of the year, they may not be accurate enough in the future Smart Grid scenarios. Accordingly, we assume that by better understanding the customers' real-time consumption needs, and by being able to partially shape these needs, the energy supplying companies could procure energy at lower prices. This motivated our work on a specific demand management concept and its enabling technology, which is centered around the system capability to detect potentials for optimization by monitoring how home devices are being used, by learning to predict when and for how long the devices will be used, and interacting with users in order to enhance the precision of predictions.

The details of the concept and its realization are presented in this paper in the following way. Section 10.2 describes the GREEN HOME Gateway architecture and functions, Sect. 10.3 briefly reviews the concept of the Demand Response, and Sect. 10.4 describes the details of the concept for Grid Responsiveness. In Sect. 10.5 the model of home activities is further described. Section 10.6 presents our forecasting approach, Sect. 10.7 the test bed suitable for the user lab-based experiments, and Sect. 10.8 provides conclusions and plans for further work.

10.2 Home Gateway Functions

The central part of the GREEN Home system is the Home Gateway that interconnects all devices within the home environment. In general, the Home Gateway platform may be distributed: computationally intensive tasks may be physically spread among several networked computation platforms, including the Cloud.

Regarding the intelligence of home devices we take into consideration that some devices may be more intelligent than others. For example, our assumption is that a simple next generation appliance will be capable of measuring its power and will offers a simple on/off switching or program switching interface; a more advanced device will be capable of getting control signals, e.g., information outlining the dynamic energy price, so that it can autonomously optimize its operation by scheduling higher load in the periods of lower prices. All home devices communicate with the Home Gateway, so that the device consumption information, as well as configuration information can be used by the home system. Appliances with no direct user interfaces may also be configured through specialized user interaction devices, e.g., a tablet or a touch screen.

User Interfaces for interaction with the system (through smart displays, pervasive interfaces, etc.) and with the next generation of appliances (embedded displays, switches, etc.) are essential components of the system. Although the system has to be able to autonomously reason and make decisions, the user must be given the means to engage in the operation of the system in order to guarantee high user satisfaction and sustainable user interest in energy management.

The role of the Home Gateway is to support many different applications and use-cases of automated energy management. In designing the GREEN HOME system component architecture we regarded requirements of a number of different home automation and energy management use-cases such as in-home or remote control, visualization for energy awareness, data sharing for social energy awareness, risk reduction via anomaly detection, emergency-signal-based demand management, price-based demand management, forecast-based demand management (the focus of this paper), market-based flexibility trading, and electric vehicle battery management. To facilitate different use-cases efficiently the architecture of the system must be flexible and based on extensible model. This motivates our approach for flexible system modeling based on the use of semantic technologies [12]. Such system realization includes ontology and the knowledge-base to organize system data. The ontology also includes the model of system alarms and is the basis for automated system actions, as we demonstrated in integration of the energy management system within an overall home autonomous system [13].

This semantic-based core is a common component of the system; for the forecast-based demand management use-case it is complemented with the following specific components:

- *The Monitoring Component* interacts with the home devices in order to acquire all relevant data. These include time series of device-based consumption, sensor data reading (e.g., temperature), device status, etc.
- *The External Data Integration Component* integrates other important data that may come from external sources such as weather predictions.
- *The System Configuration Component* elicit through interaction with users system data that cannot be obtained automatically, user profiles and preferences, and user flexibility in using devices. This information is time-stamped and saved in the knowledge-base
- *The Forecasting Component* takes as the input (1) historic per-device consumption data, (2) historic weather and holiday data, and (3) weather predictions and provides as an output a possible profile of devices' use for the next day.
- *The Plan Configuration Component* updates the profile of devices' use for the next day through interaction with users. For example the user can inform the system of his planned absence. Based on this configuration plans are prepared for each device in the home.
- *The Smart Grid Interaction Component* exchanges information with the system of the supplier. The home system receives from the supplier a price profile for the next day and provides a power usage profile based on the day-ahead updated forecast.
- *The Plan Execution Component* monitors the device usage i.e., per device power consumption against the forecasted profile and triggers the control of intelligent schedule-able devices in order to minimize the difference between the real and predicted consumption if possible. It also updates the forecast for the remaining period of the day and sends this updated forecast together with the real consumption data in 15-min periods.

10.3 Demand Response

Demand response refers to technical and business programs that are offered by energy companies, mostly utilities (DSOs), and involve customers who are willing to respond to different types of information from the utility, e.g., received via telephone calls, or specialized in-home devices, by reducing their consumption manually or automatically [14]. Many such programs have been deployed and tested in the field. The extensive research in demand response systems also resulted in a Open Automatic Demand Response standard (openADR) [15].

As compared with demand response programs typically offered by distribution companies or integrated utilities, where within a contractually defined period a number of criticality events can be received and shall result in load management actions for grid stabilization, our demand-management approach focuses on the needs of the energy suppliers in the liberalized market. In the liberalized market, energy suppliers can have customers in many different distribution areas, and are interested to win new customers based on attractive offers. They are not concerned with the grid stability, they currently do not receive consumption data of their customers in the real time, and have no access to consumption data of their potential customers. Currently they offer mostly fix prices to their customers, and the difference between this fix retail (consumer) price and the dynamic wholesale price at different markets is the source of earnings or losses. Therefore, in defining specific demand response schemes the energy suppliers can either offer dynamic prices to customers, to share the risk by directly exposing the consumer to higher prices as a motivation for cost-based load shedding or scheduling away from expensive peaks, or they can offer programs in which the responsiveness of customers help them to purchase energy at lower prices and to share this gain according to some incentive scheme.

The issue of dynamic prices has been considered in several theoretical and practical studies. In a liberalized market in which customers can select from multiple suppliers, the home energy management system should in principle be able to receive dynamic prices from multiple suppliers so that the customer could reduce the consumption when the price is high but also dynamically select the best offer in real-time. Our focus in this work is not on selection among multi-supplier offers, but we developed a demand management scheme which exploits an existing contract between the customer and the supplier and aims at getting incentives (monetary compensations) by deploying a home energy management system with a forecasting and automated load scheduling capability.

Our scheme is essentially a decision-support scheme as the Home Gateway supports users in discovering possibilities for consumption optimization and in communicating forecasts of their consumption to their supplier.

10.4 Grid Responsiveness Concept

The GREEN HOME concept of responsiveness for consumption optimization is designed also to meet requirements regarding privacy-preservation and scalability of operation. Both requirements are addressed by carefully discriminating among the information that can be collected within the home, and the information to be exchanged among the home and the supplier domain. As already mentioned, we anticipate a near-term vision in which all appliances in the home are capable of monitoring their own consumption and communicating with the Home Gateway. The users may use some devices more flexibly that the others. For these, the users can directly input their preferences regarding the device scheduling or regarding the device priority. This flexibility is the basis for consumption optimization through automated scheduling. In addition the sensors in home will be able to monitor the temperature, humidity, and light levels. In our concept these data are basis for building the knowledge about the users' behavior. This very detailed knowledge is only accessible for the in-home control and forecasting function. It does not leave the home and can be secured within the home domain.

The information that leaves the home domain is the aggregated consumption information – both the real-consumption data and the forecasted consumption. Especially forecasting data require very careful handling, in the first place through encrypted transport. Forecasts of future consumption may be quite accurate (for example a user can announce his absence in the next week) and can be of great value for the supplier company. On the other hand if intercepted it may be a source of security risks, and if changed it may incur damage as wrong information would be used on the supplier side to make purchases.

The communication for demand management among the home domain and the supplier domain happens periodically and we differentiate between the day-ahead forecast-plan exchange and the intra-day consumption and forecast exchange. In the day-ahead exchange the supplier domain system first requests the forecast from the GREEN Home system on the previous day. It provides an energy consumption preference curve (proportional to forecasted prices) to guide possible scheduling.

The home system replies with a day-ahead forecast reflecting demand flexibility, and the supplier system completes the interaction with the agreed plan. The day-ahead forecast-plan exchange includes direct interaction with users; the intra-day forecast involves only respective computational intelligence at the Home Gateway and the supplier side system. The details are provided in the following sub-sections.

10.4.1 The Day-Ahead Exchange

Once per day a gateway receives the preference curve from the supplier with the values between 0 (to be avoided) and 1 (preferred) for each minute of the next day. The home system generates a forecast of the schedules of the home devices' use. Applying a probabilistic approach the forecasting component selects from a number of schedules of different probability the one with the highest probability, and presents it to the user. The user interface for this interaction enables the user to select from several options:

- Holiday; the user specify that the next day (or several next days) the home will not be occupied and that only always-on devices or specially scheduled devices (e.g., time-scheduled lights, HVAC or water heater scheduled for the hours before the residents arrival) shall be switched on. In this case it could be expected that the uncertainty is quite low.
- Active day; the user provides feedback about the forecasted schedule, deleting the devices that will not be used and selecting the devices that will be used. The user also provides scheduling constraints, e.g., the washing machine will

wash for 4 h (based on the selected program) and shall be started after 17:00 and shall finish before 3:00 the other day.

- Passive day; the user provides no feedback to the schedule. In this case the uncertainty is quite high and there is no flexibility.

After the interaction with the user is completed, the forecasting component schedules flexible devices according to the preference curve and sends the updated plan to the supplier system encoded as a Forecast.xml document. Within the offered flexibility the supplier selects one specific option and creates a Plan.xml document that describes the load profile which the supplier is now expecting from the user. The Home Gateway will make device activating decisions based on this selected plan.

10.4.2 The Intra-day Exchange

In the intra-day operation, there is an additional exchange of information among the home domain and the supplier domain which happens each 15 min. The Plan Execution Component of the Home Gateway monitors the energy consumption according to the agreed plan. The system detects that there are some deviations from the forecasted schedule (e.g. some devices are switched on although this was not planned) and so the total consumption for this period may be different from the planned one. Nevertheless, because the system has the knowledge of these events it can update prediction regarding the use of energy in the next period. This update of prediction together with the consumption information is sent to the supplier system.

We selected the 15 min time interval here because the smart meter regulation prescribes that meter readings are acquired each 15 min. It should be noted that currently reading the meter is in the domain of the distribution system operator. The readings are collected only for the accounting purposes and presented to respective suppliers e.g., once per month. As contrary to this approach, we are designing the system in which supplier collects consumption and forecasting data from many homes in the real-time (15 min) in order to be able to make better intra-day trading decisions.

The scheduling of flexible devices is performed with the goal to minimize the deviation of the real consumption from the selected plan. The flexibility includes switching on schedule-able appliances or switching on/off any of interruptible active appliances.

10.4.3 User Responsiveness

The major benefit for the energy supplier company introduced by this program is that it acquires not only the real-time (15 min interval) consumption data but also consumption predictions which though not fully accurate, much better represent the reality than standard profiles that are used today. The supplier can also suggest how to use flexibility in scheduling. In particular when the GREEN Home resident uses options such as "holiday" or "active day" and updates the forecasts to better represent her plans and actual activities, the benefits for the supplier may be quite high.

The system at home can also learn about user responsiveness – the willingness to use active day and holiday options, the quality of his forecast updates, and also about the predictability or regularity in user life. This information may be quite important for the supplier as regularity in home activities may better qualify users for such a program.

10.5   The Model of Home Activities

The proposed forecasting-based demand management incorporates learning and requires specific data from the past. For the evaluation purposes where such data is not available we designed a simulator [16] which implements a flexible model of home and user activities and generates a number of related time series: (a) for each device a synthetic consumption profile with power level reading for each minute, (b) for each resident the presence profile with the values 1 (present at home) and 0 (absent from home) for each minute (c) outdoors temperature per each minute (d) the light condition with values 1 (high level) and 0 (low level) (e) the time of year is captured in a model where for each day we can have the following values: work day, weekend, holiday,

The model of home activities is based on a simple ontology which captures the house, devices, residents, temperature, light conditions, and seasonality. Regarding temperature the simulator use real temperature data, e.g., we used the historical meteorological data for Vienna. The real temperature data is a time series of daily measurements including two readings for each day: the maximum and the minimum temperature. Based on these readings we created synthetic temperature levels for each minute created random values uniformly distributed within the minimum and maximum level. The outdoors light conditions are also real data; here we use the sunrise and the sunset data for Vienna, and assign the value 0 to the interval between the sunset and the sunrise, and the value 1 for the interval between the sunrise and the sunset. The temperature data are used to model the heating and cooling function. In our simple model we included a function which defines several intervals of temperature and for each interval defines the number of (electricity-based) heating or cooling hours requires. Once the number of hours is determined for each day, based on previously generated per-minute temperature values the heating or cooling are scheduled to start in the minutes of the day at which the triggering temperatures are for the first time reached.

All appliances are described with their power use. In the simulator we assume constant power profiles for the switched-on state. Each lamp in the home is also classified as the one that is dependent/not dependent on the light levels. For all lamps that are dependent on light levels the percentage of switch-on time is specified.

Hence all such sources of light will be assumed to be switched on after the sunset and their active period will be distributed between the sunset and the sunrise.

The model defines that home is used by a family which consists of several residents; the home includes devices which implement functions. For each day the user can be present or absent depending on the type of the day (weekend, working day, holiday). The family as a whole can be absent from home, these are the days which are labeled as big holiday. Each of the residents in home is characterized with his profile of interactions with devices. For each user, personalized intervals for daily activities can be specified and for each of these, the devices that user can select to use, the average time of using devices and probability of using these devices. These profiles are used to generate per-device usage data.

Using presented models the simulator generates different time series which are the input for the forecasting function.

10.6 Forecasting of the Uncertainty Level

Forecasting of consumption uncertainty and flexibility intervals is the main component of the Home Gateway intelligence. In fact, the gateway generates potential schedules of device usage and predicts resident's presence based on the model constructed through learning, first within some specific learning period in which the gateway do not participate in any demand response activity. After this period of learning the forecasting system can start to participate in the demand response program and continues to learn through monitoring and interactions with users.

The forecasting component predicts for the next day, for each device (which is not always on) the amount of time that it will be used. We tested and integrated an approaches based on the similar day model and statistic method that detects the frequency of the patterns of device use, similar to [17]. The forecasting takes into account the status of each devices, the presence of residents, the hour of the day, the day of the week, and the month of the year information, the holiday information, the temperature interval information, and the light condition information.

10.7 The Test Bed

To evaluate proposed concept we designed and implemented a test-bed that integrates both emulated and simulated functions [18]. The demonstrator setup integrates a ONENET [19]-based network with tablet PCs emulating home devices (see Fig. 10.2 for a screenshot), hence real-time aspects of in-home communication between appliances and the Gateway can be studied. Novel UIs for the device control are running as applications on the tablet PCs. The set-up also includes a network of smart plugs [20]. The input for a test run in a test-bed includes time series of per-device energy consumption generated by the simulator, e.g., modeling

10 GREEN HOME: The Concept and Study of Grid Responsiveness

Fig. 10.2 A screenshot of the energy consumption simulator's main menu that shows all devices that can be simulated (TV set, washing machine, refrigerator, rechargers, etc.). Each simulation runs in real time, therefore real time values are sent to the Gateway using ONENET via an attached communication node. If the simulator is not connected to a real network of devices, it offers a much faster simulation mode, too, in order to easily see consequences of events

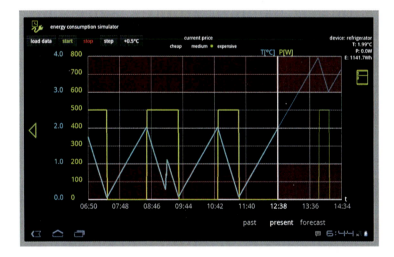

Fig. 10.3 Simulator screenshot: A refrigerator is simulated. The current temperature inside the refrigerator is drawn in *blue*, the energy consumption is drawn in *yellow*, the *pale* part in the right shows the current forecast (i.e. the screenshot shows the state at time 12:38). As soon as the temperature reaches an upper threshold (here 2.0°C) a new cooling cycle begins, in this case having a predefined power consumption of 500 W. The "current price" was just set to "expensive", therefore the algorithm tries to delay the cooling cycle as long as possible by allowing an upper threshold of 4.0°C, in order to consume less energy. The short peak at around 09:15 comes from a simulated opening of the refrigerator door, which leads to a quick temperature increase

Fig. 10.4 Screenshot of a washing machine simulation with a typical load profile. The *yellow curve* is the objected load profile (corresponding to the chosen washing programme, in this case "colored laundry 60 degrees") and the *red curve* is the envelope function showing the maximum energy consumption allowed (by the Gateway) at all times. The numbers stated at the load profile curve depict minutes the point of the profile can be delayed without influencing the washing process (e.g. the start of the spin cycle might be delayed a few minutes). Current simulated time is 06:34 (i.e. all data shown to the right of this vertical bar is a forecast). 07:27 is the configured latest allowed end for the completion of the washing programme

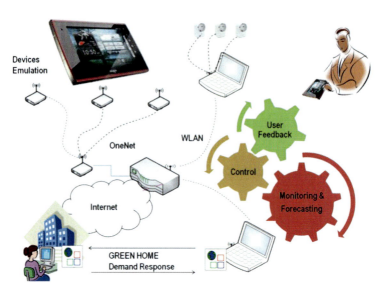

Fig. 10.5 GREEN Home test bed architecture

a consumption of the last 3 years. These are used for training the forecasting model. One additional year is simulated to represent the real usage. The test environment supports all interactions according to the described approach. The test starts with the day-ahead forecast and user interaction for updates. The test UI includes the representation of the emulated real consumption, and the representation of the forecasted one (see Figs. 10.3 and 10.4). The user can update the forecast (according to real consumption and real presence) or leave it as it is. The user can also select passive day option. The automatic monitoring and forecasting interactions (in the real-world performed each 15 min) are performed in much shorter periods. For evaluation of performance the system calculates the difference between the (updated) forecast plan and the real consumption.

The aim of the test bed study is to validate our concept in several scenarios of interest in particular with respect to user responses to different user interface options.

The test bed is schematically illustrated in Fig. 10.5.

10.8 Conclusions

In this paper we presented a concept, an implementation and an evaluation environment for a collaborative energy management using the model of home responsiveness based on an in-home consumption forecasting. This new model utilizes continuous interactions between the energy management systems in the home domain and in the energy supplying company domain. Information exchanged includes higher precision forecasts generated within the home domain. These forecasts are calculated by the home energy management system which collects device data and infers user behavior. The protocol is implemented using Web-service approach, exploiting the benefits of ubiquitously supported and standardized technology.

Our future work will focus evaluations and system extensions. In user studies we plan to evaluate the acceptance of the concept and user interfaces involved. Particularly interesting is how user can control the precision of forecasts that are sent to the provider by his own actions. For example if user is willing to disclose that he is out of his home for several days he can achieve higher benefit because of better forecast. On the other hand he may not be willing to do so. The second venue of future work is to quantify benefits of our approach in large scale scenarios. Again the knowledge of the users' privacy preferences will be incorporated in these studies. The third challenge is that of full distribution of home management intelligence. In this scenario the schedule for appliances is not calculated centrally at the gateway but is a result of a distributed algorithm. The appliances have a higher level of autonomy and do collaborative allocation of available energy.

Acknowledgements The work was supported within a research project Grid Responsive Energy Efficient Networked Home (GREEN HOME).The Telecommunications Research Center Vienna (FTW) is supported by the Austrian government and the City of Vienna within the competence center program COMET.

References

1. Taylor, A.S., Harper, R., Swan, L., Izadi, S., Sellen, A., Perry, M.: Homes that make us smart. Pers. Ubiquitous Comput. **11**(5), 383–393 (2007)
2. Tomic, S., Fensel, A.V., Schwanzer, M., Kojic Veljovic, M., Stefanovic, M.: Semantics for energy efficiency. In: Applied Semantic Technologies: Using Semantics in Intelligent Information Processing. CRC Press, Taylor and Francis, USA (2011)
3. Schmidt, A., van Laerhoven, K.: How to build smart appliances? IEEE Pers. Commun. **8**(4), 66–71 (2001)
4. Open Service Gateway Initiative, Device Expert Group. http://www.osgi.org. Accessed August 2013
5. GSMA: Vision of Smart Home, The Role of Mobile in the Home of the Future, Sept 2011. Available from: http://www.gsmaembeddedmobile.comasVisionofSmartHomeReport.pdf
6. State-of-the-art of energy management, e-Health and community-service requirements on common service delivery frameworks, Project FIGARO Future Internet Gateway-based Architecture of Residential Networks, Mar 2011. Available from: http://www.ict-figaro.euasFIGARO-Deliverable-5-1.pdf
7. CENELEC: Smart House Roadmap. Available from: ftp://ftp.cenorm.be/CENELEC/SmartHouseasSmartHouseRoadmap.pdf. October 2010
8. Standardization Trends of OSGi Technology, NTT Technical Review, 6(1), (2008). Available from: https://www.ntt-review.jp
9. Requirements for the Software Modularity on the Home Gateway, HGI, 2011. http://www.homegatewayinitiative.org/publis/RD-008-R3.pdf
10. Use Cases and Architecture for a Home Energy Management Service, HGI, 2011. Available from: http://www.homegatewayinitiative.orgasuse-cases-and-architecture-for-home-energy-Management-service.pdf
11. EnWG -Weimar, Lastprofile und Einspeiseprofile, August 2013, available at http://www.enwg-weimar.de/tech-strom-lastprofile.php
12. Tomic, S., Fensel, A.V., Pellegrini, T.: SESAME Demonstrator: Ontologies, services and policies for energy efficiency. In: Proceedings of the I-SEMANTICS, Graz, Sept 2010. ACM.
13. Next Generation Home, GENIO, 2011. http://www.celtic-initiative.org/projects/genio
14. Albadi, M.H., El-Saadany, E.F.: Demand Response in electricity markets: An Overview, Power Engineering Society General Meeting, 2007. IEEE , pp. 1–5, 24–28 June 2007
15. Open ADR: http://drrc.lbl.gov/openadr. Accessed August 2013
16. Wenninger, J., Haase, J.: Efficient building automation simulation using system on chip simulation techniques. In: Proceedings of 2013 26th International Conference on Architecture of Computing Systems (ARCS), Prague (2013)
17. Barbato, A., Capone, A., Rodolfi, M., Tagliaferri, D.: Forecasting the usage of household appliances through power meter sensors for demand management in the smart grid. In: IEEE International Conference on Smart Grid Communications (SmartGridComm), Brussels, Belgium, 2011, pp. 404, 409, 17–20 Oct 2011
18. Rathmair, M., Haase, J.: Simulator for smart load management in home appliances. In: The Fourth International Conference on Advances in System Simulation (SIMUL), Lisbon, Portugal, pp. 1–6, Nov 2012
19. ONE NET Protocol: http://www.one-net.info. Accessed August 2013
20. PlugWise: http://www.plugwise.com. Accessed August 2013

Chapter 11
Polynomial Metamodel-Based Fast Optimization of Nanoscale PLL Components

Saraju P. Mohanty and Elias Kougianos

Abstract As the complexity of nanoscale-CMOS analog/mixed-signal (AMS) circuits and systems grows, the challenges of their design becomes exponentially more difficult. Performing accurate design simulations that entail exhaustive design space exploration has become infeasible with the increasing complexity of nano-CMOS circuits and systems integration, coupled with aggressive scaling of process technologies. Transistor-level SPICE simulations with full parasitics (RCLK) of complex circuits, which provide silicon accurate results, have run times in the order of days or weeks. With ever shrinking time to market pressures, the simulation time proves to be impractical as it can lead to longer design cycle times. The simulation time factor is further aggravated by additional design and process parameters which have to be accounted for due to increased sensitivity in deeply scaled technologies. In order to mitigate this problem, this chapter presents a two-stage approach that uses layout-accurate metamodels and efficient search algorithms for fast mixed-signal circuit and system optimization. The different components of a Phase-Locked Loop (PLL) are considered as a case study. First, the metamodel creation process is presented. A simulated annealing based optimization algorithm is then discussed for power optimization of the PLL components. It is shown that the metamodel approach speeds up the optimization phase by 2,000× with very good accuracy. The power consumption of the circuit is decreased by 22% for the baseline design and is within 8% of the circuit netlist-based, but computationally expensive approach.

S.P. Mohanty (✉)
Computer Science and Engineering, University of North Texas, Denton, TX 76203, USA
e-mail: saraju.mohanty@unt.edu

E. Kougianos
Engineering Technology, University of North Texas, Denton, TX 76203, USA
e-mail: eliask@unt.edu

J. Haase (ed.), *Models, Methods, and Tools for Complex Chip Design*, Lecture Notes in Electrical Engineering 265, DOI 10.1007/978-3-319-01418-0_11,
© Springer International Publishing Switzerland 2014

11.1 Introduction

The market desire for smaller and yet more powerful computing devices has been the driving force towards more complex integrated designs for analog/mixed-signal (AMS) circuits and systems. With technology scaling into the deep nanometer region, the number of design and process parameters to be taken into consideration increases significantly. AMS component design is a complex and time consuming process especially at the optimization and physical design stages [13]. In addition, the presence of parasitics after the layout stage has a very dramatic effect on the output, hence making traditional numerical simulation methods inefficient [23]. This in turn increases the computation time for accurate exploration of the design space.

Even with advanced computing resources, the full simulation of complex circuits takes days and sometimes weeks to complete. With an increased number of parameters and pronounced effects from process variation, efficient and accurate design optimization becomes prohibitive due to the expensive simulation costs. The inherent problem is one of computational efficiency versus accuracy. The scaling of technology deeper into nanometer regions aggravates this design problem. Previously proposed methods to increase the efficiency by reducing the computation time include the use of interpolating functions and fast optimization algorithms for design optimization [22]. Thus, there is a pressing need for fast and accurate design flows and optimization approaches for mixed-signal circuit and system design exploration.

The mitigation of AMS design optimization problems can be achieved by one or more of the following approaches: (1) Reduction of the simulation time; (2) Reduction of the optimization time; (3) Reduction of the number of layout iterations needed. The research introduced in this paper proposes a new design flow for fast and accurate optimization of complex mixed-signal circuits and systems. The proposed design flow crates and uses accurate and fast metamodels of the actual circuit and performs optimization on the metamodels, not the actual circuit. The key concept of the metamodel assisted optimization is depicted in Fig. 11.1.

Metamodels (which are models of models) are used in many different fields to simplify the design process, especially when the sampling of the design space for optimization is very costly or time consuming [9, 11]. Metamodels are also known as surrogate models. Using mathematical functions or algorithms in the optimization step speeds up the process since each iteration of the calculations does not require analog simulation of the circuit or recreation of the physical design. In this chapter we address the power metric to simplify and explain the metamodel creation process. Creating a metamodel is a very crucial step, since the manufacturing price of the circuit is very high and it is essential to produce the most accurate metamodel possible, given a fixed simulation time budget. Phased locked loop components are investigated separately and their metamodeling is presented. The metamodels are then used to optimize the individual circuits.

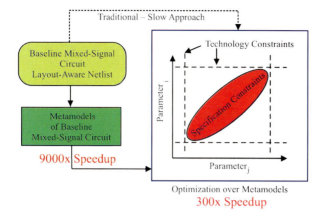

Fig. 11.1 The proposed metamodel-assisted optimization

The *novel contributions of this chapter* to the state-of-art are the following:

1. A metamodel assisted fast AMS design flow that can significantly reduce design cycle time.
2. An approach for polynomial metamodel generation which can be used for any technology node.
3. A simulated annealing based algorithm to optimize different components of a PLL for power (including leakage).
4. A 22% decrease in power is achieved when using the metamodeling approach to the initial baseline design and 8% decrease in power is reached over the traditional netlist-based approach.
5. Metamodeling is 2,000× faster compared to the netlist-based optimization.
6. A metamodeling MATLAB GUI toolbox is presented to assist designers in metamodel creation, verification and optimization.

The rest of the chapter is organized as follows: Sect. 11.2 discusses the novel proposed flow. A discussion of related prior literature is included in Sect. 11.3. An overview of a PLL and its different components is presented in Sect. 11.4. Section 11.5 explains the steps needed to create an accurate metamodel. Section 11.6 presents the optimization algorithm used. Section 11.7 presents the experimental results. Summary, conclusion and future research are discussed in Sect. 11.8.

11.2 Proposed Novel Fast Analog/Mixed-Signal Design Flow

The proposed novel fast design flow for AMS circuit and system optimization is depicted in Fig. 11.2. The proposed overall fast AMS design optimization flow can be logically divided into four steps as follows: (1) Baseline AMS component

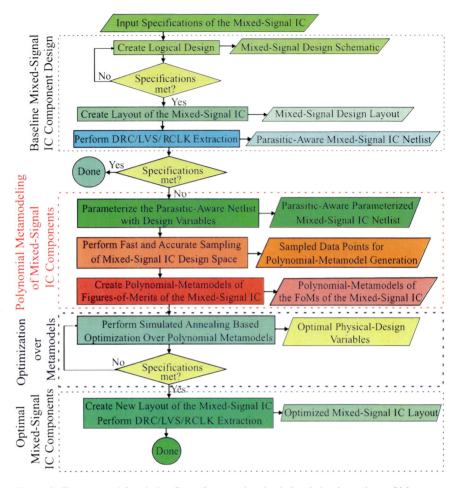

Fig. 11.2 The proposed fast design flow of nanoscale mixed-signal circuits such as a PLL

design and characterization, (2) Polynomial metamodel generation, (3) Optimization over metamodels, and (4) Optimal AMS component design and characterization. Speedup in the proposed metamodel may come from many aspects: (1) Use of optimization iterations over the metamodels instead of the netlist. (2) Decoupling of the optimization from an analog simulation (SPICE) framework. (3) Use of faster optimization algorithms. (4) Use of exactly two manual layout steps.

In the first phase of the novel flow, the baseline physical design of AMS components is performed. At this phase minor optimization may be performed over the schematic designs which are fast to run since the netlists do not contain parasitics. At this point a baseline physical design of AMS components is made and the baseline circuit SPICE netlist with parasitics (RCLK, R – Resistance, C – Capacitance, L – Self Inductance and K – Mutual Inductance) is extracted from

11 Polynomial Metamodel-Based Fast Optimization of Nanoscale PLL Components

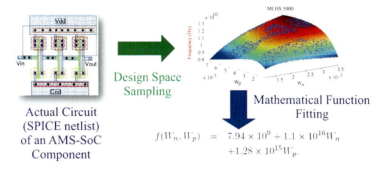

Fig. 11.3 The concept of metamodel: transforming circuit netlist to mathematical expressions

the physical design. The parasitic aware netlist is then parameterized in terms of the design variables. The polynomial metamodel generation phase uses a fast sampling algorithm (for example Latin Hypercube Sampling or LHS) of the AMS component design space. Polynomial metamodels are generated from these sparse sampled data for each of the target Figures-of-Merit (FoMs) of the PLL components. At the optimization phase of the proposed flow, the well-established Simulated Annealing Algorithm (SAA) is used to optimize the FOMs. The specification constraints are set by using one or more metamodels and the main FoM is then optimized within the constraints. One of the main benefits of using the metamodels is that they are reusable, as long as the FoMs and the process constraints are selected judiciously by the designer. Once the metamodels are created, the specification constraints can be adjusted and the optimization time using metamodels is very short in comparison to the same type of optimization done via simulations on the original model (SPICE netlist). Therefore a more dynamic design can be achieved using metamodels without being impacted by simulation time. Once the optimal design parameters are obtained from the optimization, a final (or 2nd) physical design is performed and characterized.

The speedup in the proposed flow relies heavily on the metamodels. A metamodel is a predictive mathematical algorithm or function for a given FoM such as power, frequency, jitter, leakage, and phase noise [11]. Each circuit can obviously have more than one metamodel associated with it if the optimization is multi-objective. The concept of metamodel is presented in Fig. 11.3 [12]. In essence, metamodels transform the circuit characteristics to set of equations, functions, or algorithms and decouple the circuit simulation from the continuous-time analog domain (SPICE). Any general modeling language, such as C/C++, Verilog-AMS/VHDL-AMS and Matlab/Simulink can then be used for design exploration. For the metamodel to be effectively useful, it needs to have the following properties [15]: (1) *Accuracy* – This is the capability of the metamodel to predict the system response over the entire allowable design space. (2) *Efficiency* – Efficiency is an indicator of the computational effort required for constructing the metamodel. (3) *Transparency* – This is the capability of the metamodel to provide the information concerning contributions and variations of design variables and correlation among the variables. (4) *Simplicity* – Metamodel generation should require minimal user input and should be easily adaptable to different problems and circuits. More information on metamodeling and computer experiments is given from a generic point of applicability in [9], but VLSI areas of application are not covered there.

11.3 Related Prior Research

Design space exploration approaches from high-level descriptions of analog circuits are given in [7]. The use of neural networks in the automatic synthesis of OP-AMPs is explored in [30]. RF-specific transistor sizing with explicit parasitic estimates is given in [2]. A layout-aware modeling approach for analog synthesis is given in [24] and [25]. Posynomial modeling for gate sizing is presented in [26, 27].

In [28], a surrogate modeling approach for expensive circuit-level simulation is presented that uses support vector machine (SVM)-based machine learning. In [16], the authors propose the use of metamodels for creating an inductor for CMOS circuits. The technique that the author proposes does not use sampling techniques but rather uses mathematical formulas for the model estimation and optimization. Metamodeling has been applied on IP reuse for SoC integration and microprocessor design in [18]. This approach covers a higher level design flow, is purely digital and does not create metamodels specifically for CMOS circuits. In [31], a surrogate modeling approach is also used for statistical wire-length estimation. Metamodeling has been implemented for grid computing in [14].

An order reduction technique, macromodeling is discussed in [1, 3, 4, 6]. Some authors use the terms "macromodel" and "metamodel" interchangeably but they are very distinct. A macromodel is simply a reduced complexity (order) representation of the circuit but is still a netlist, necessitating the use of an analog (SPICE) simulator. On the other hand, a metamodel is a language and simulator independent model of the original model (hence the term meta). In this context, a VCO parametric metamodeling approach is given in [8].

11.4 Design of PLL Component Circuits

As a specific case study of the proposed mixed-signal system design flow the design of a Phase Locked Loop (PLL) is considered. A PLL is a closed loop feedback control system that consists of a Phase Detector (PD), Charge Pump (CP), Voltage Controlled Oscillator (e.g. LC-VCO), and Frequency Divider (FD). Thus the PLL is an excellent example of a mixed-signal system. The block diagram representation of a PLL is presented in Fig. 11.4. The following sections briefly detail each component of the PLL and highlight their design parameter complexity. For brevity, selected

11 Polynomial Metamodel-Based Fast Optimization of Nanoscale PLL Components

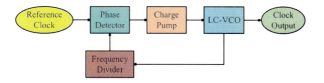

Fig. 11.4 Block diagram of a Phase Locked Loop (PLL)

layouts are provided for an 180 nm CMOS process. In this research, the PLL's output has a 2.25 GHz frequency for wireless local area network (WLAN) applications.

11.4.1 Phase Detector

The phase detector is an important component of the PLL. Its action enables the phase differences in the loop to be detected and the resultant error voltage to be produced. A proportional phase detector directs the charge pump to supply charge amounts in proportion to the phase error detected. Phase detectors range from a very simple XOR gate to complex logic circuits consisting of flip-flops. The schematic and physical design of the phase detector for 180 nm CMOS is provided in Fig. 11.5. This circuit has many transistors and it is not efficient to consider them all separately in the design exploration process. In order to simplify the optimization process, all transistor widths in one flip flop are assigned to design parameters W_{nDFF1} and W_{pDFF1} for NMOS and PMOS, respectively. In the second flip flop they are W_{nDFF2} and W_{pDFF2}, respectively. For the remaining NMOS and PMOS they are W_n and W_p, respectively.

11.4.2 Loop Filter and Charge Pump

The charge pump provides stabilization of spurious fluctuation of currents and the switching time in order to minimize the spurs in the VCO input [10]. The output signal from the charge pump is applied to the loop filter which determines the PLL's dynamic characteristics. A low-pass RC filter is used that passes frequency signals within the range of the VCO. The loop filter can affect tracking and capture ranges and maximum slew rate. For metamodel generation and optimization, the width of all charge pump transistors are considered.

11.4.3 LC Voltage Controlled Oscillator

An LC-VCO is an electronic oscillator specifically designed to be controlled by a voltage input. The operating frequency of the LC-VCO can be mainly controlled

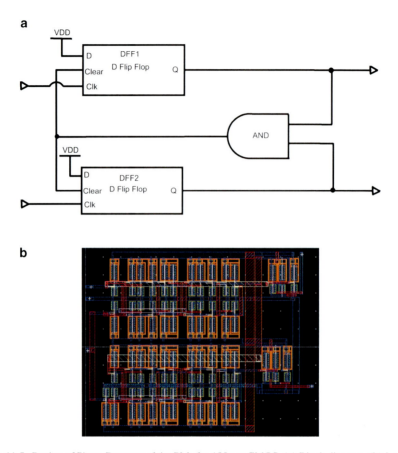

Fig. 11.5 Design of Phase Detector of the PLL for 180 nm CMOS. (**a**) Block diagram. (**b**) Layout

by applying a DC input voltage. Since the PLL's output is mostly dependent on the output of the LC-VCO, this work targets a 2.25 GHz frequency for WLAN applications. The LC-VCO has only two easily adjustable parameters other than the size of the LC tank which is considered fixed. Since it is a symmetric circuit the two NMOS and two PMOS transistor widths have been selected as design variables. The layout of the LC-VCO is shown in Fig. 11.6.

11.4.4 Frequency Divider

The basic operation of the frequency divider is to reduce the frequency of a continuous train of pulse waveforms fed to it as an input signal, to approximately half. The frequency divider is implemented using true single phase logic. The baseline frequency divider was designed using the minimal width values allowed

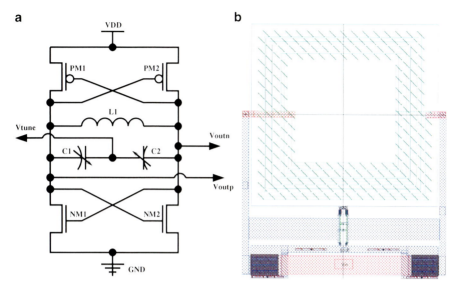

Fig. 11.6 LC-VCO design for 180 nm CMOS. (**a**) Schematic. (**b**) Layout

by the technology node: 400 nm for the NMOS and 800 nm for PMOS. For each transistor in the divider, the width of each NMOS and each PMOS is varied and is considered as a design parameter separately. Therefore the divider circuit has nine design parameters. The boundaries were set to 400 nm $\leq (W_n$ and $W_p) \leq 1{,}600$ nm, where W_n and W_p are the widths of NMOS and PMOS respectively, while the nominal values have been selected to be $W_n = 400$ nm and $W_p = 800$ nm.

11.5 Proposed Approach for Generation of Fast and Layout-Accurate Metamodels

The creation of metamodels is the crucial step in this proposed fast design flow. Selecting the right prediction function that has low error variance compared to the actual simulated analysis is crucial. Since it is not possible to exhaustively sample the design space when a large number of degrees of freedom (design variables) is present, the accuracy of the model varies based on the form of the metamodel. To decrease the complexity of metamodel creation this research targets the generation of polynomial functions due to the small number of variables for each subcircuit. This process can be extended to other functions such as splines and artificial neural networks, which can be generated from the same sample data. To make this process robust the following flow for metamodel creation has been created and described in Algorithm 6. In this algorithm, ERROR refers to the error between the predicted value at a sample point and the actual result from the netlist simulation.

> **Algorithm 6:** Layout-accurate polynomial metamodel generation
>
> 1: Obtain the parasitic aware netlist of an AMS component.
> 2: Parameterize the netlist for target tuning variables.
> 3: Sample the AMS component design space; sample ← generate_sample_data().
> 4: Verify the sample data; verify ← generate_verify_data().
> 5: Perform mean data centering; centered ← center(sample).
> 6: **for** degree ← 1 to p **do**
> 7:   sample_mm(degree) ← stepwise(sample, degree).
> 8:   centered_mm(degree) ← stepwise(centered, degree).
> 9:   sample_ERROR(degree) = verify(sample_mm, verify).
> 10:   centered_ERROR(degree) ← verify(centered_mm, verify).
> 11: **end for**
> 12: pick_lowest(sample_ERROR, centered_ERROR).

The final outcome is a multivariate polynomial function of degree p in the n design variables $\mathbf{x} = x_1, x_2, \cdots, x_n$ for the predicted response $\hat{y}$:

$$\hat{y}(\mathbf{x}) = \sum_{|\alpha| \leq p} \mathbf{c}_\alpha \mathbf{x}^\alpha, \tag{11.1}$$

where the multi-index $\alpha = (\alpha_1, \alpha_2, \cdots, \alpha_n)$ takes values in $\mathbb{N}_0^n$, $|\alpha| \equiv \alpha_1 + \alpha_2 + \cdots + \alpha_n$, $\mathbf{x}^\alpha \equiv (x^{\alpha_1}, x^{\alpha_2}, \cdots, x^{\alpha_n})$ and $\mathbf{c}_\alpha \equiv (c_{\alpha_1}, c_{\alpha_2}, \cdots, c_{\alpha_n})$ are the undetermined fitting coefficients. Since $0 \leq |\alpha| \leq p$, the total number of such terms is:

$$N(p,n) = \sum_{k=0}^{p} \binom{n+k-1}{k} \tag{11.2}$$

As the number of design variables (n) and/or the degree of the multinomial (p) increase, $N(p,n)$ increases exponentially.

The final metamodel degree p is selected so that it provides the least error to the data points and the lowest error for verification points. As the error metric value for fitted data points shows the error for points that are being sampled, it may not always be a good metric to use since the function can be possibly over fitted, i.e. it can pass through all sample points but can have a very large error between samples. Points in between the sampled data are used to verify the accuracy of the function and based on the error metric value calculated from that step, the metamodel function degree is selected. Algorithm 6 generated the best fitted polynomial function within the given criteria. If the accuracy still does not satisfy the specifications, the number of samples has to be increased to generate more fitting functions. The initial number of samples can be quite small to reduce the sampling time and can be gradually increased for more accuracy. It is a matter of experience to select appropriate minimal number of sample points needed for accuracy and speed trade-offs.

11.5.1 *Data Sampling*

The starting step for creating the metamodel is to perform simulations for a sample of the design space, determined by the available simulation budget. There are many sampling options available including Monte Carlo, Latin Hypercubes, Middle Latin Hypercubes, and Design of Experiments [11]. In this paper all metamodels have been generated using Middle Latin Hypercube Sampling (MLHS) [17] which divides the design space into Latin Hypercubes [29] and then samples the middle point of each hypercube. Our previous research has shown that data is distributed more evenly, even though MLHS does not sample the edges where the parameters have minimum and maximum values, but due to the polynomial function behavior it is compensated by the slope of the function as long as the behavior of the sample data is not sporadic at the edges [11]. All metamodels have been generated by 1,000 MLHS samples and 100 MLHS samples were used for verification.

A matrix L is generated to code each row as a term in the polynomial power for each parameter except when the value is 0 as it is implied that the term is not present. As an example, for second order code with two parameters, L has the following form:

$$L = \begin{bmatrix} 0 & 0 \\ 1 & 0 \\ 2 & 0 \\ 0 & 1 \\ 0 & 2 \\ 1 & 1 \end{bmatrix} \tag{11.3}$$

Let us assume that matrix $X \equiv [x_1, x_2]$ where x_1, x_2 are design parameters. Then the design matrix template (DMT) for X using L code becomes $DMT(X,L) = [0, x_1, x_1^2, x_2, x_2^2, x_1 x_2]$. The design matrix (DM) is created by using MLHS of the x_1 and x_2 parameters, resulting in a matrix of the following form:

$$DM^T(X,L) = \begin{bmatrix} 0 & \cdots & 0 \\ \text{mlhs}(x_1)_1 & \cdots & \text{mlhs}(x_1)_n \\ \text{mlhs}(x_1)_1^2 & \cdots & \text{mlhs}(x_1)_n^2 \\ \text{mlhs}(x_2)_1 & \cdots & \text{mlhs}(x_2)_n \\ \text{mlhs}(x_2)_1^2 & \cdots & \text{mlhs}(x_2)_n^2 \\ \text{mlhs}(x_1)_1 \text{mlhs}(x_2)_1 & \cdots & \text{mlhs}(x_1)_n \text{mlhs}(x_2)_n \end{bmatrix} \tag{11.4}$$

The corresponding Y matrix is then sampled from the original design and is based on the parameters x_1 and x_2 for each row in the DM matrix.

11.5.2 Data Centering

Mean data centering is used to alleviate abrupt changes in the data. It centers and scales each column of X before fitting. Z_{score} for each parameter in the circuit is calculated by using the following expression:

$$Z_{score} = \left(\frac{x_i - \mu_i}{\sigma_i}\right), \tag{11.5}$$

where x_i is the i-th design parameter, μ_i is the mean of the data points for that parameter, and σ_i is its standard deviation. This approach is mostly used when the parameter range is very high and the data is not distributed evenly.

11.5.3 Stepwise Regression

Stepwise regression is conducted on the sample data to create the polynomial of first to sixth order [20]. The resulting polynomial may not include all terms since stepwise regression iteratively removes coefficients that are not statistically significant, resulting in metamodels with fewer coefficients without losing accuracy. This method may build different models from the same set of potential terms depending on the terms included in the initial model and the order in which terms are moved in and out. The function stepwise(), which is embedded in MATLAB, tests different initial models and outputs the coefficients for the best model for that order.

11.5.4 Verification of the Metamodel

For the estimation of the accuracy of the metamodels, several error metrics can be used [15]. Three important error metrics for metamodel performance analysis are the following: (1) Root-Mean Square Error (RMSE), (2) Relative Average Absolute Error (RAAE), and (3) R-Square (R^2). RMSE represents the departure of metamodel from the real response (SPICE parasitic netlist results are considered "golden"). RMSE is calculated in the following manner:

$$RMSE = \sqrt{\left(\frac{1}{N}\right) \sum_{k=1}^{N} (y(x_k) - \hat{y}(x_k))^2}, \tag{11.6}$$

where N is the number of sample points, $y(x_k)$ is the metamodel response at the k-th sample point and $\hat{y}(x_k)$ is the "golden" response. A smaller RMSE means that the metamodel is more accurate. The RAEE is calculated in the following manner:

11 Polynomial Metamodel-Based Fast Optimization of Nanoscale PLL Components

Table 11.1 Metamodel polynomial degree comparison for PLL component power consumption

PLL components	Degree	R^2	R^2 adjusted	Coefficients total	Coefficients in model	RMSE fit (W)	RMSE verification (W)
	1	0.8610	0.9997	7	3	2.1×10^{-10}	6.6×10^{-10}
	2	0.9603	0.9993	28	18	1.1×10^{-10}	7.1×10^{-10}
Phase	3	0.9663	0.9989	84	34	1.0×10^{-10}	7.0×10^{-10}
Detector	4	0.9748	0.9974	210	94	9.2×10^{-11}	7.1×10^{-10}
	5	0.9926	0.9955	462	206	8.2×10^{-11}	7.1×10^{-10}
	6	0.9968	0.9938	924	662	5.4×10^{-11}	9.7×10^{-10}
	1	0.9905	1	5	5	2.3×10^{-6}	8.4×10^{-4}
	2	0.9968	1	15	14	1.3×10^{-6}	8.4×10^{-4}
Charge	3	0.9987	1	35	31	8.6×10^{-7}	8.5×10^{-4}
Pump	4	0.9995	1	70	54	5.3×10^{-7}	8.4×10^{-4}
	5	0.9999	1	126	87	2.8×10^{-7}	8.2×10^{-4}
	6	1	1	210	151	1.4×10^{-7}	8.3×10^{-4}
	1	0.9237	0.9985	3	3	4.5×10^{-5}	4.1×10^{-5}
	2	0.9769	0.9999	6	6	2.5×10^{-5}	2.1×10^{-5}
LC-VCO	3	0.9878	0.9999	11	11	1.8×10^{-5}	1.7×10^{-5}
	4	0.9927	0.9999	16	16	1.4×10^{-5}	1.4×10^{-5}
	5	0.9942	0.9999	21	20	1.2×10^{-5}	1.2×10^{-5}
	6	0.9946	0.9999	28	20	1.2×10^{-5}	1.1×10^{-5}
	1	0.2954	0.9943	10	9	7.6×10^{-6}	7.7×10^{-6}
	2	0.3503	0.9901	55	16	7.3×10^{-6}	5.9×10^{-6}
Divider	3	0.4504	0.9618	220	66	6.9×10^{-6}	5.8×10^{-6}
	4	0.6787	0.8828	715	268	6.0×10^{-6}	9.1×10^{-6}
	5	1	1	2,002	999	0	1.1×10^{-4}

$$RAEE = \left(\frac{\sum_{k=1}^{N} |y(x_k) - \hat{y}(x_k)|}{N \times \sigma} \right), \tag{11.7}$$

where σ is the standard deviation of all error terms. A smaller RAAE means more accurate metamodel. R^2 is calculated in the following manner:

$$R^2 = \left(1 - \left(\frac{1}{N \times \sigma^2} \right) \sum_{k=1}^{N} (y(x_k) - \hat{y}(x_k))^2 \right). \tag{11.8}$$

A larger R^2 means a more accurate metamodel. In this chapter RMSE and R^2 have been used as error accuracy measures.

The RMSE value that is calculated from metamodel fitting for sampling points is not a good metric for the metamodel fit. Extra verification points are required to ensure that the fit is also good at other points than the ones being sampled. Verification samples are checked to make sure that the data points are not the same as the sampling points. The lowest RMSE value for different metamodel functions is then selected as the best function. Polynomial functions of order 1 through 6 are considered in this research. A total of 12 metamodels for the power consumption of each component, 6 metamodels using the sample points with data centering and 6 metamodels without data centering are generated and then compared. The lowest RMSE verification value is used to pick the most accurate metamodel. Table 11.1 compares the various metamodels.

11.6 Proposed Metamodel Based Design Optimization

For the optimization of the PLL components over the polynomial metamodels many heuristic algorithms can be used. The optimization over metamodels instead of the SPICE netslists allows accurate and otherwise computationally expensive algorithms to iterate extensively before converging to an optimal solution. We propose the use of the well established simulated annealing algorithm in this research. The average power dissipation is considered as the objective for minimization.

Simulated annealing optimization is an extension of the Monte Carlo algorithm and simulates the annealing process which is used in metallurgy. This gives the simulated annealing algorithm random characteristics. More details of the theory

---

Algorithm 7: Simulated annealing optimization for P_n parameters

1: Initialize iteration counter: $Counter \leftarrow 0$.
2: Initialize first feasible solution $S_i \leftarrow mid(P_n)$ for each parameter P.
3: Determine initial $Power_i$ using polynomial metamodels for the solution S_i.
4: Initialize temperature T and cooling rate α_T.
5: **while** ($\Delta_{Power}! = 0$) **do**
6:     $Counter \leftarrow$ Maximum number of iterations.
7:     **while** ($Counter > 0$) **do**
8:         Generate random transition from S_i to S_i^*.
9:         **if** (S_i^* is acceptable solution) **then**
10:             Update the results; $result \leftarrow S_i^*$.
11:             break both while loops.
12:         **else**
13:             Calculate average power of AMS component from polynomial metamodels for solution S_i as $Power_i$.
14:             Calculate average power of AMS component from polynomial metamodels for solution S_j as $Power_j^*$.
15:             Calculate the difference in power dissipation as: $\Delta_{Power} \leftarrow Power_i - Power_j^*$
16:             **if** $\left(\Delta_{Power} < 0 \ random(0,1) < e^{\left(\frac{\Delta Power}{T} \right)} \right)$ **then**
17:                 Update the solution; $S \leftarrow S_i^*$.
18:             **end if**
19:         **end if**
20:         Decrement the counter; $Counter \leftarrow Counter - 1$.
21:     **end while**
22:     Change the temperature; $T \leftarrow \alpha_T \times T$.
23: **end while**
24: **return** $result$.

---

11 Polynomial Metamodel-Based Fast Optimization of Nanoscale PLL Components

behind the simulated annealing algorithm can be found in [5]. This paper explored the algorithm for circuit optimization as the number of parameters can be quite large. By its nature, the simulated annealing algorithm has a random component and two successive runs can produce different results. The steps of simulated annealing based search are presented in Algorithm 7.

The algorithm takes random walks through the design space starting from its middle point, looking for points with low energies. Parameter S calculations are based on the polynomial metamodels of the FoMs which consider the PLL frequency as a constraint, and returns the component's power value if the design is within the frequency specifications. If the design is not within the frequency specifications the returned value is high, making the algorithm ignore the parameter values of that step. In each step, the probability of taking a step is determined by the Boltzmann distribution, $p = \left(e^{\frac{\Delta_{Power}}{T}}\right)$ if Δ_{Power} is high, and $p = 1$ when Δ_{Power} is low. Therefore a step will occur if a new value is better than the previous one. If the new value is worse, the transition can still occur, and its likelihood is proportional to the temperature T and inversely proportional to Δ_{Power}.

11.7 Experimental Results

The algorithm implementations and simulations were performed in an integrated environment of Cadence and MATLAB. A GUI is developed for easy use of the design flow, as shown in Fig. 11.7. A complete development of this AMS

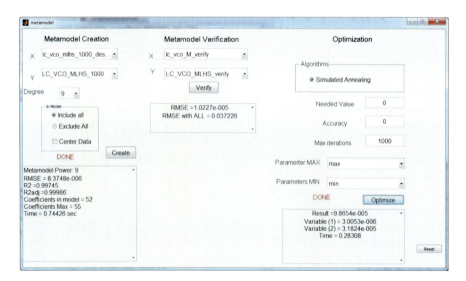

Fig. 11.7 Screen shot of metamodeling toolbox which will be released for educational usage once development is completed

Table 11.2 Multiobjective optimization using power and frequency metamodels for LC-VCO with 1% accuracy for 2.25 GHz

	Power	Frequency	Error to verification	Error to target
Prediction	0.153 mW	2.23 Ghz	0.047 mW	0.02 Ghz
Verification	0.110 mW	2.25 Ghz	–	0 Ghz
Polynomial order	4	5		

Table 11.3 Power metamodel versus netlist circuit optimization comparison

Circuit (parameters)	Metamodel			Netlist	
	Order	Prediction (W)	Time (s)	(W)	Time (min)
Phase Detector (6)	1	7.35×10^{-9}	0.07	6.81×10^{-9}	≈ 19
Charge Pump (4) (without data centering)	2	3.01×10^{-5}	0.11	3.11×10^{-5}	≈ 18
LC-VCO (2)	5	8.84×10^{-5}	1.22	9.54×10^{-5}	≈ 18
Divider (1)	3	1.20×10^{-5}	1.05	3.43×10^{-5}	≈ 8

design framework will eventually become a comprehensive automatic exploration environment.

The results for PLL component multiobjective optimization which has been performed using two metamodels, one for power and one for frequency, are shown in Table 11.2. The frequency metamodel is used as constraint in the optimization process and is set to within 1% of 2.25 GHz. The simulated annealing algorithm is used to minimize the power with the given constraint. As a result, the chosen metamodels are of order 4 and 5 for power and frequency respectfully. The accuracy of the metamodel prediction values are verified using SPICE simulations and show that the metamodel error to actual simulation at the near optimal point is 47 μW for power and 20 MHz for frequency. It is also interesting to note that the verification simulation achieved closer results to the target values with lower power and frequency being directly on target.

The results for the different PLL components are shown in Table 11.3. The metamodel generation time is generally in the order of seconds and depends on the maximum number of coefficients that can be present in the model, since the stepwise regression process is very much dependent on the number of coefficients. Higher order polynomials add more calculation time, but still the largest 6th order polynomial with 9 parameters took roughly 2 min to create. The optimization time is also in seconds in comparison to simulation optimization.

The final results for power consumption for all PLL components are shown in Table 11.4. The initial power has been measured at the lowest minimal values for each parameter considered in all circuits. The order of each metamodel varies and was selected from those having the lowest RMSE values. It is easy to see that on average the metamodeling approach has reached better results than the simulation approach, even though simulation optimization provides better results for the divider.

Table 11.4 Power optimization using metamodels for different components

PLL components	Phase detector (nW)	Charge pump/ loop filter (µW)	LC-VCO (µW)	Divider (µW)	Total power (µW)	CPU time
Initial power	9.31	21.84	75.5	60.5	157.9	
Metamodel optimization	6.80	3.10	71.4	48.1	122.6	7.48 s
Simulation optimization	6.87	3.11	95.4	34.3	132.8	81 min

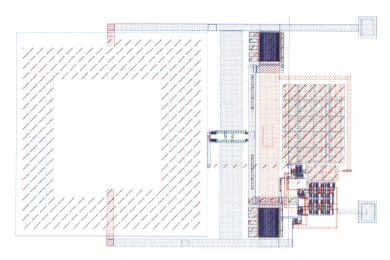

Fig. 11.8 Layout of the 180 nm CMOS based complete PLL

Once optimization of different PLL components is performed, the overall PLL is designed by combining them. The complete physical design of the PLL for 180 nm CMOS is shown in Fig. 11.8. Subsequent characterization of the PLL finds the following specifications: 0.92 mW average power (including leakage), 2.25 GHz center frequency, and $525 \times 326\,\mu m^2$ silicon area.

11.8 Summary, Conclusions, and Future Direction of Research

From this research we can conclude that the second order polynomial, which forms the basis for Response Surface Methodologies (RSM), does not always accurately capture the complexity of the response in a multidimensional design space. For a PLL circuit, higher order polynomials provide better results. The metamodeling design flow lowers the design optimization phase by roughly 2,000× compared to the simulation based optimization approach if used for IP reuse. The metamodel generation is the slowest step of the proposed design process and provides speedup that is roughly equal to the ratio of the number of samples and netlist optimization sampling. One can argue that the metamodeling approach sampling and model generation stage is time consuming, but considering that the metamodels are reusable and can be used for circuit verification in addition to optimization, this approach becomes very attractive for AMS SoC design and verification.

It may be noted that metamodeling techniques have been well researched in many disciplines [15]. The research is in full swing for its applicability in VLSI design automation. With the increasing complexity and dimensionality of

11 Polynomial Metamodel-Based Fast Optimization of Nanoscale PLL Components

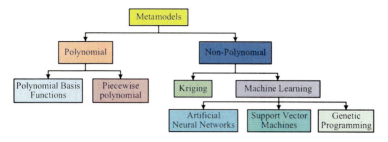

Fig. 11.9 Different types of metamodels

Table 11.5 Metamodeling comparative perspective for a PLL with center frequency f_c of 2.7 GHz

PLL	ANN metamodel		Polynomial metamodel	
FoM	RMSE	Time to create	RMSE	Time to create
Center Frequency	48 MHz (1.8% of f_c)	11 h for sampling +10 min for training	78 MHz (2.9% of f_c)	11 h for sampling +1 min for creation

designs, the need for more efficient and accurate metamodeling techniques grows. A classification and analysis of metamodeling techniques is presented in Fig. 11.9 [19, 32]. Various metamodeling approaches, including second order polynomial techniques, Multivariate Adaptive Regression Splines (MARS), boosting, Support Vector Machine (SVM), Artificial Neural Networks (ANN), Genetic Programming (GP), and Kriging are being investigated for accuracy and speed trade-offs.

For a specific comparison, ANN and polynomial metamodels are considered and presented in Table 11.5. For design optimization over the actual circuit netlist, 100 iterations take 10,000 min, which is 7 days. This is the worst case analysis. On the other hand, both ANN and metamodels reduce that time significantly. The RMSE for polynomial metamodel is larger than the ANN metamodels. However, polynomial metamodels are straight forward to create and do not need sophisticated training as is the case of ANN metamodels.

In the future, the research problem of metamodeling and design optimization over metamodels can be tackled in multiple fronts. The sampling phase is the most time consuming phase of metamodeling. What is the best alternative to Monte Carlo for large circuits to capture the complete design space accurately is still an open question. Higher parameter circuits and more circuit characterization will be considered in our future research. The correlations among parameters in affecting a target FoM needs more research. Capturing the process variations accurately through metamodeling is essential for nanoscale circuit and system optimization. Also, multiobjective optimization functions involving more than one, possibly conflicting FoMs, is a future research direction.

Acknowledgements The chapter is based on the following presentation [21]. The authors would like to acknowledge the help of UNT graduate Dr. Oleg Garitselov and Mr. J. M. Molina who presented the conference version. The authors would like to thank the Editors of this book and FDL 2012 [21] organizers.

References

1. Agarwal, A., Vemuri, R.: Hierarchical performance macromodels of feasible regions for synthesis of analog and RF circuits. In: IEEE/ACM International Conference on Computer-Aided Design, San Jose, pp. 430–436 (2005)
2. Agarwal, A., Vemuri, R.: Layout-aware RF circuit synthesis driven by worst case parasitic corners. In: 2005 IEEE International Conference on Computer Design: VLSI in Computers and Processors, San Jose (2005)
3. Agarwal, A., Wolfe, G., Vemuri, R.: Accuracy driven performance macromodeling of feasible regions during synthesis of analog circuits. In: Proceedings of the 15th ACM Great Lakes Symposium on VLSI, Chicago, pp. 482–487 (2005)
4. Basu, S., Kommineni, B., Vemuri, R.: Variation-aware macromodeling and synthesis of analog circuits using spline center and range method and dynamically reduced design space. In: 22nd International Conference on VLSI Design, New Delhi, pp. 433–438 (2009)
5. Bertsimas, D., Tsitsiklis, J.: Simulated annealing. Stat. Sci. **8**(1), 10–15 (1993)
6. Ding, M., Vemuri, R.: Efficient analog performance macromodeling via sequential design space decomposition. In: 19th International Conference on VLSI Design, Hyderabad, p. 4. (2006)
7. Doboli, A., Vemuri, R.: Exploration-based high-level synthesis of linear analog systems operating at low/medium frequencies. IEEE Trans. Comput. Aided Des. Integr. Circuits Syst. **22**(11), 1556–1568 (2003)
8. Dong, W., Feng, Z., Li, P.: Efficient VCO phase macromodel generation considering statistical parametric variations. In: Proceedings of the IEEE/ACM International Conference on Computer-Aided Design, San Jose, pp. 874–878 (2007)
9. Fan, K.T., Li, R., Sudjianto, A.: Design and Modeling for Computer Experiments. Chapman and Hall/CRC, Boca Raton (2006)
10. Gardner, F.: Charge-pump phase-lock loops. IEEE Trans. Commun. **28**(11), 1849–1858 (1980). [legacy, pre-1988]
11. Garitselov, O., Mohanty, S.P., Kougianos, E.: A comparative study of metamodels for fast and accurate simulation of nano-CMOS circuits. IEEE Trans. Semicond. Manuf. **25**(1), 26–36 (2012)
12. Garitselov, O., Mohanty, S.P., Kougianos, E.: Accurate polynomial metamodeling-based ultra-fast bee colony optimization of a nano-CMOS phase-locked loop. J. Low Power Electron. **8**(3), 317–328 (2012)
13. Ghai, D., Mohanty, S.P., Kougianos, E.: Design of parasitic and process-variation aware nano-CMOS RF circuits: a VCO case study. IEEE Trans. VLSI Syst. **17**(9), 1339–1342 (2009)
14. Hendrickx, W., Gorissen, D., Dhaene, T.: Grid enabled sequential design and adaptive metamodeling. In: Proceedings of the Winter Simulation Conference, Monterey, pp. 872–881 (2006)
15. Jin, R., Chen, W., Simpson, T.W.: Comparative studies of metamodelling techniques under multiple modelling criteria. Struct. Multidiscip. Optim. **23**, 1–13 (2001)
16. Lamecki, A., Balewski, L., Mrozowski, M.: Towards automated full-wave design of microwave circuits. In: 17th International Conference on Microwaves, Radar and Wireless Communications, Wroclaw, pp. 1–2 (2008)
17. Lesh, F.H.: Multi-dimensional least-squares polynomial curve fitting. Commun. ACM **2**, 29–30 (1959)

18. Mathaikutty, D.A., Shukla, S.: Metamodeling driven IP reuse for system-on-chip integration and microprocessor design. Artech House, Norwood, MA 02062 USA (2007)
19. McConaghy, T., Gielen, G.: Analysis of simulation-driven numerical performance modeling techniques for application to analog circuit optimization. In: Proceedings of the IEEE International Symposium on Circuits and Systems, (ISCAS), Iasi, vol. 2, pp. 1298–1301 (2005)
20. McCray, A.T., McNames, J., Abercrombie, D.: Stepwise regression for identifying sources of variation in a semiconductor manufacturing process. In: IEEE Conference and Workshop on Advanced Semiconductor Manufacturing, Boston, pp. 448–452 (2004)
21. Mohanty, S.P., Kougianos, E., Garitselov, O., Molina, J.M.: Polynomial-metamodel assisted fast power optimization of nano-CMOS PLL components. In: Proceeding of the 2012 Forum on Specification and Design Languages, Vienna, pp. 233–238 (2012)
22. Mohanty, S.P., Kougianos, E., Okobiah, O.: Optimal design of a dual-oxide nano-CMOS universal level converter for multi-v dd socs. Analog Integr. Circuits Signal Process. **72**(2), 451–467 (2012)
23. Park, J., Choi, K., Allstot, D.J.: Parasitic-aware design and optimization of a fully integrated CMOS wideband amplifier. In: Proceedings of the 8th Asia South Pacific Design Automation Conference, Kitakyushu, pp. 904–907 (2003)
24. Pradhan, A., Vemuri, R.: A layout-aware analog synthesis procedure inclusive of dynamic module geometry selection. In: Proceedings of the 18th ACM Great Lakes Symposium on VLSI, Orlando, pp. 159–162 (2008)
25. Pradhan, A., Vemuri, R.: Efficient synthesis of a uniformly spread layout aware pareto surface for analog circuits. In: Proceedings of the 22nd International Conference on VLSI Design, New Delhi, pp. 131–136 (2009)
26. Roy, S., Chen, C.C.P.: Smartsmooth: a linear time convexity preserving smoothing algorithm for numerically convex data with application to VLSI design. In: Asia and South Pacific Design Automation Conference, Yokohama, pp. 559–564 (2007)
27. Roy, S., Chen, W., Chung-Ping Chen, C., Hu, Y.H.: Numerically convex forms and their application in gate sizing. IEEE Trans. Comput. Aided Des. Integr. Circuits Syst. **26**(9), 1637–1647 (2007)
28. Samanta, R., Hu, J., Li, P.: Discrete buffer and wire sizing for link-based non-tree clock networks. IEEE Trans. Very Large Scale Integr. (VLSI) Syst. **18**(7), 1025–1035 (2010)
29. Tang, B.: Orthogonal array-based latin hypercubes. J. Am. Stat. Assoc. **88**(424), 1392–1397 (1993)
30. Wolfe, G., Vemuri, R.: Extraction and use of neural network models in automated synthesis of operational amplifiers. IEEE Trans. Comput. Aided Des. Integr. Circuits Syst. **22**(2), 198–212 (2003)
31. Wong, J.L., Davoodi, A., Khanderwal, A., Srivastava, A., Potkonjak, M.: A statistical methodology for wire-length prediction. IEEE Trans. Comput. Aided Des. Integr. Circuits Syst. **25**(7), 1327–1336 (2006)
32. Yelten, M.B., Zhu, T., Koziel, S., Franzon, P.D., Steer, M.: Demystifying surrogate modeling for circuits and systems. IEEE Circuits Syst. Mag. **12**(1), 45–63 (2012)

Chapter 12
Methodology and Example-Driven Interconnect Synthesis for Designing Heterogeneous Coarse-Grain Reconfigurable Architectures

Johann Glaser and Clifford Wolf

Abstract Low power consumption or high execution speed is achieved by making an application specific design. However, today's systems also require flexibility in order to allow running similar or updated applications (e.g. due to changing standards). Finding a good trade-off between reconfigurability and performance is a challenge.

This work presents a design methodology to generate application-domain specific heterogeneous coarse-grain reconfigurable architectures. The specification of the reconfigurable architecture is given by a set of example applications which define the whole range of its required functionality. These applications are analyzed to extract common building blocks, which can be reused between them.

In the next step, the circuits of the application are merged to a single reconfigurable module. The major part of this work describes the according tool and its algorithm. Its main task is to optimize the interconnect by hierarchically grouping the functional units. Additional resources can be added to enable future applications. The tool generates the HDL source for a module with the instances of all blocks and the reconfigurable interconnect. The feasibility of the methodology is demonstrated by the design of reconfigurable architectures for digital filters as well as simple logic networks.

12.1 Introduction

In current system design a shift to employ reconfigurable logic tries to utilize their benefits for various applications. Typical wireless sensor network (WSN) nodes are supplied from batteries or utilize energy haversting. Therefore the main goal

---

J. Glaser (✉) • C. Wolf
Institute for Computer Technology, Vienna University of Technology, Vienna, Austria
e-mail: glaser@ict.tuwien.ac.at; clifford@clifford.at

J. Haase (ed.), *Models, Methods, and Tools for Complex Chip Design*, Lecture Notes
in Electrical Engineering 265, DOI 10.1007/978-3-319-01418-0_12,
© Springer International Publishing Switzerland 2014

is to optimize a WSN node for ultra low power consumption. Unfortunately, the CPU as main controller consumes power even for very simple tasks. By adding a dedicated reconfigurable hardware module to offload the CPU for such simple tasks as sensor measurements or network MAC layer handling, a large reduction in the power consumption can be achieved [4]. These reconfigurable modules also enable the use of the SoC in multiple different environments, thus sharing the non-recurring engineering (NRE) costs.

Accelerators for computer vision systems should support various algorithms. Currently this is achieved by implementing all algorithms in parallel and switching between them. Since the algorithms also have common operations, a reconfigurable system can reduce the required hardware resources. In multi-standard and multi-function communication systems the same approach leads to a reduction of hardware resources [12].

Reconfigurable logic is classified by its granularity. The widely used FPGAs are fine-grained and pose a large overhead in terms of area and power. This is avoided by coarse-grained reconfigurable systems that achieve an ASIC-like performance at much lower power consumption and chip area [11, 15]. For the above mentioned applications, domain-specific reconfigurable circuits with heterogeneous, tailored blocks and a non-regular interconnection can provide further reduction in power and area [7].

In this work, a methodology for the design of heterogeneous coarse-grain reconfigurable circuits is presented. From a set of different actual applications, the set of required (possibly reconfigurable) hardware blocks and the interconnect between them is deduced. The grouping of the blocks is optimized to minimize the hardware resources of the interconnect.

This work is an extended version of [13]. First we review the design and usage of custom reconfigurable hardware. Then a detailed view on the design methodology is given. This is followed by a review and evaluation of interconnect topologies. The main part of this work is an optimization algorithm for the automatic synthesis of this interconnect. Then a short section introduces a feature-rich Verilog synthesis tool which is used for design entry of the presented methodology. This is followed by an evaluation of the algorithm results. The work ends with conclusions and future work.

12.2 Development of Reconfigurable Hardware

The generation of reconfigurable circuits is split in two phases. In the so called "*pre-silicon phase*" the reconfigurable hardware structures are designed for the application class. Secondly, in the "*post-silicon phase*" the reconfigurable silicon circuit is used to implement the actual application [5, 7].

In this work an approach is presented that provides the (semi-) automated generation of the pre-silicon circuit and can generate the configuration data for an actual application in the post-silicon phase.

12.2.1 Pre-silicon Phase

In the pre-silicon phase, the reconfigurable circuit is designed. As first step, its specification is derived from the set of (usually similar) actual applications, which will be implemented in the reconfigurable logic. During this design space exploration, the "needs of [the] applications [...] drive the construction of the fabric" [11, p. 1]. This approach requires the a-priori knowledge of all future applications and it is generally not possible to implement a different application with the resulting fabric. To enable yet unknown applications, we propose oversizing, i.e., to include additional hardware and interconnect resources into the fabric.

The specification includes information on the employed blocks (also called functional units or cells) (e.g. adders, FSMs, ...), which can be reconfigurable themselves (e.g. an adder be reconfigured as a subtracter, reconfigurable FSM [6]). Additionally it includes the number of instances of each block as well as details on the connections among them.

12.2.2 Post-silicon Phase

In the post-silicon phase after production the actual application has to be implemented by configuring the silicon structure designed in the pre-silicon phase. So, on the one hand, the post-silicon phase is limited by the results of the pre-silicon phase. On the other hand, the pre-silicon phase requires information on the actual implementations later used in the post-silicon phase to provide the required resources.

12.3 Design Methodology

The reconfigurable module as the result of the pre-silicon development phase will be integrated into the whole SoC. Therefore the resulting design data has to be compatible with an industry-standard ASIC design flow. This is best accomplished by delivering the reconfigurable module as a soft IP core. The required deliverables include structural and RTL (register-transfer level) hardware description (e.g., VHDL, Verilog) as well as guidelines and constraints for synthesis and place and route. Additionally, information and tools for the post-silicon phase have to be provided.

The design of such reconfigurable module soft IP cores should be assisted and automated by dedicated tools. This requires a systematic design approach which will be described in this section.

12.3.1 Specification

The start of the development requires a precise specification of the reconfigurable module. The application class of the module only coarsely defines the functionality. However, the interfaces of the reconfigurable modules to other modules of the SoC and outside the SoC can be derived. On the other hand, the functionality and inventory of the reconfigurable module itself can be specified in two different ways:

1. Define the functionality and the required flexibility in abstract terms.
2. Use a set of concrete applications, which define the whole range of the required functionality of the reconfigurable module.

This work only deals with the second form of specification. The design of a reconfigurable module is thus broken down to first developing a set of concrete applications. Then, from this set one reconfigurable circuit is derived which is able to implement each of these concrete applications.

The specification of each concrete, i.e., example application has to be easy to translate to a logic netlist to facilitate further processing by automated tools. It should employ an existing type of hardware description so that the designers do not have to learn a new one. Finally, the description should be supported by industry-standard verification tools to achieve a first-time-right SoC design. All these requirements are fulfilled by common hardware description languages like VHDL and Verilog.

12.3.2 Application Analysis

The first step of the development of a reconfigurable module is to develop and verify a set of example applications (see "App"s in Fig. 12.1). In the second step, these are processed to derive the inventory of the reconfigurable module. This consists of a pool of coarse-grain cells (e.g., FSMs, adders, ...) (which might be reconfigurable themselves) plus a reconfigurable interconnect for flexible connections among them [7, 13] (compare Fig. 12.3).

For the processing of the example applications, a special coarse-grain synthesis tool creates a netlist representation of each application. These netlists are analyzed to extract coarse-grain cells and to find commonalities between all example applications (see top "Synthesis" box in Fig. 12.1). Commonly used cells are candidates for reuse in the reconfigurable module. Analogously to the FSM plus datapath (FSM + D) concept, the control logic of each example application is mapped to an FSM using FSM extraction while all data processing is implemented using dedicated coarse-grain cells.

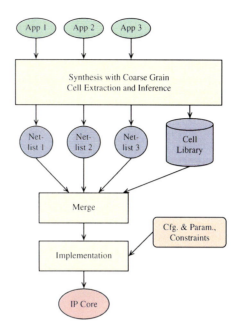

Fig. 12.1 Graphical representation of the design methodology (description is given in the text)

While FSMs are very generic building blocks, which easily can be implemented with a reconfigurable cell, e.g. as a TR-FSM [6], the development of a reconfigurable datapath is a more complex task. At the beginning of the development, a mostly manual approach is employed. The coarse-grain synthesis tool creates netlists with instances of simple coarse-grain cells (e.g., adders, shifters, ...). The designer has to investigate the schematic and manually identify related cells which together build a larger, more complex coarse-grain cell (e.g., calculating the absolute difference of two numbers, a counter, ...). He can also generalize such groups to configurable multi-function cells (e.g. an adder and subtracter or a whole ALU). Further more, using frequent subgraph mining [12] allows to automatically identify sub-circuits which are common to multiple example applications.

For each such coarse-grain cell a small module is designed (again in a HDL). These together build the cell library (see Fig. 12.1). We call this semi-automatic procedure "coarse-grain cell extraction". It is the core point for optimization of the final reconfigurable module.

When the cell library for the reconfigurable module is finished, all example applications are analyzed again with "coarse-grain cell inference". This uses subgraph isomorphism to identify and replace all sub-circuits in each example application by coarse-grain cells from the cell library. For the further processing, each example application has to be described by a netlist of only FSMs and instances of cells from the cell library and connections among them (see "Netlist"s in Fig. 12.1).

12.3.3 Merge

After the application analysis, all separate example applications are merged to a single common reconfigurable module, which can implement each of the example applications (see Fig. 12.1). Therefore all used coarse-grain cells from the cell library including reconfigurable FSMs have to be instantiated in the appropriate number. To increase the flexibility of the final reconfigurable module, additional instances of critical cells can be added. Finally, a flexible and reconfigurable interconnect is created and optimized.

The result of this step is an RTL representation of the reconfigurable module plus meta-information for the post-silicon phase to specify how to setup the configuration data.

12.3.4 Implementation

To complete the reconfigurable module, storage and interfaces for configuration and parameterization are added (see Fig. 12.1). Together with guidelines and constraints for synthesis and place and route, the reconfigurable module is provided as soft IP core. It can be integrated into the whole SoC and processed with a standard ASIC design flow.

12.3.5 Verification

Verification is a major concern in ASIC development, therefore the described development methodology provides full coverage from the example applications to the finished IP core. Firstly, the functionality of each example application (see "App"s in Fig. 12.1) can be verified by simulation as well as formal verification due to the choice of common HDLs for specification.

Secondly, the individual netlists created by coarse-grain cell inference can be checked for equivalence to the original HDL. Since these netlists might already contain reconfigurable cells (e.g., FSMs), the configuration values have to be set to the proper values using according commands of the equivalence checking tools. The simulation of the netlist simply uses the original testbench but requires the application of configuration values before start.

Finally, it is also possible to verify whether the final reconfigurable module implements any given example application. Again, the according original testbench is employed and the required configuration data has to be applied before the start of the simulation. Additionally, equivalence checking can be employed to verify the logical equivalence of the (appropriately configured) reconfigurable module to each example application.

12.3.6 Post-silicon Phase

In the previous sub-sections, it was assumed that the configuration data for each example application is generated together with the application analysis and merging steps. Therefore no dedicated post-silicon phase is required for these applications. However, for a new application a full post-silicon design phase is required. This is similar to designing FPGA applications, although the results of the pre-silicon phase limit the design space of these new applications.

In a first step, the new application is synthesized and coarse-grain cell inference is performed. Then, similar to the merging step above, the netlist is mapped to the inventory of the reconfigurable module and the signals are routed through the interconnect. Finally, the configuration data is generated to setup the cells and the interconnect. The result can be verified using simulation and equivalence checking.

12.3.7 Tools

The above described design methodology requires tools to assist the designer. For the synthesis of the example applications, a flexible and customizable coarse-grain synthesis tool is required. This will be described in short in Sect. 12.7. However, the major part of this paper is on the tool to merge the netlists of the example applications and generate and optimize the interconnect (Sects. 12.4–12.6).

12.4 Interconnect for Reconfigurable Modules

Most applications of coarse-grain reconfigurable logic are designed for computational tasks [15]. These use an array of homogeneous functional units connected with a highly regular interconnect (e.g. mesh structure), similar to FPGAs. In contrast, the presented approach assumes heterogeneous functional units (cell types), which also require a non-regular interconnect.

12.4.1 Common Topologies

Different interconnect topologies are evaluated in this section. The most powerful topology provides connections from every output to all inputs. The disadvantages are a large circuit overhead. On the other hand, a minimalistic interconnect with a small number of multiplexers to switch between alternative datapaths (compare [12]) does not allow to implement yet-unknown applications in the reconfigurable circuit.

Mesh structures are an alternative to the layered topology, but also assume homogeneous FUs that can be configured to perform each of its basic functionalities. The interconnect itself requires a high number of switches which pose a high overhead in terms of silicon area and power.

In SoCs, a bus topology is used to connect the CPU with the memory and all peripherals. For reconfigurable logic circuits with all cells working in parallel, this leads to high traffic and thus congestions [15]. The utilization of every cell is reduced and the total processing time protracted, which is not acceptable in the domain of low-power circuits.

A tree based interconnect topology [10] allows to group the cells to provide short paths through lower levels of the tree for connections, which are used frequently by the different applications. On the other hand, connections to other nodes are still possible using higher hierarchical levels of the tree. This provides a large optimization potential to reduce circuit overhead but still results in a rich set of routing resources.

12.4.2 A Tree Topology

For the implementation of the reconfigurable modules the tree topology was chosen to connect the individual cells. First, a few terms have to be defined. The circuit is built out of multiple *cells*, which are instances of various *cell types* (previously called blocks, e.g. adder, FSM, look-up tables). Each has a number of input and output ports.

Analogous to the separation of the control logic and the data-path in the FSM + D concept, each port of the cell types implements a *connection type*, e.g. bit-wide, word-wide or other categories. The connection types are defined based on compatible signaling (e.g. identical bit width) as well as semantics (e.g. clock enable vs. other control signal).

All cells are connected using a reconfigurable *interconnect*. For every connection type a separate interconnect is implemented (see Fig. 12.2) which provides connections between all ports of its connection type.

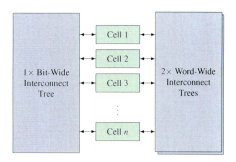

Fig. 12.2 Example interconnect with two different interconnect types (bit-wide and word-wide). The word-wide interconnect is implemented as two parallel trees

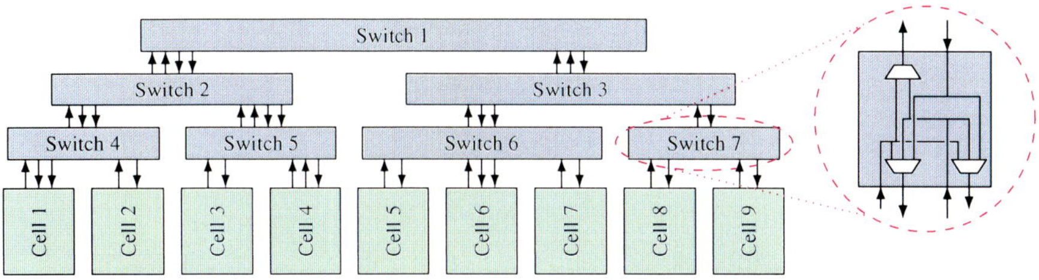

Fig. 12.3 Example interconnect with seven switches in three levels connecting nine cells of varying cell types

In the post-silicon phase an actual application is implemented by connecting the cells as given by the netlist. This specifies *nodes* of certain cell types, which are *mapped* to the cells of the reconfigurable circuit. The ports of these nodes are connected with *nets* which are routed via the interconnect of the according connection type by setting the proper configuration.

The interconnect is a tree (see Fig. 12.3) with the cells as leaf nodes and reconfigurable *switches* as inner nodes as well as the edges as connections (electrical nets).

The switches are unidirectional circuits that can be configured to connect any input port to any output port (see the detail in Fig. 12.3). The *degree* of a switch is the number of its *children*, (e.g. in Fig. 12.3, Switch 3 has a degree of two, Switch 6 has a degree of three). Each cell and each switch have a *parent* switch, except the top-most *root switch*. The *height* of the tree is the number of levels (e.g., Fig. 12.3 has a height of three).

The *routing length* of a net is the number of switches it passes from its source cell to its destination cell. The *total routing length* is the sum for all nets of a given netlist.

Each non-root switch in the tree has a number of connections to and from its parent switch. Only the number of these connections limits the capability of the interconnect to implement different netlists. Each switch can drive all outputs from any input, with one exception: A signal driven by one switch to another switch cannot be routed back to its originating switch.

To improve the connectivity, for each connection type multiple parallel trees with identical topology can be implemented (as also implemented by Ferreira et al. [3], compare Fig. 12.4 and the two word-wide interconnect trees on the right side in Fig. 12.2). Each cell is assigned to a (generally different) leaf node in each tree. Therefore each net can be routed in any tree. As each cell might be assigned to a different leaf node in each tree, the routing length of a net can be small in one tree but high in the other trees.

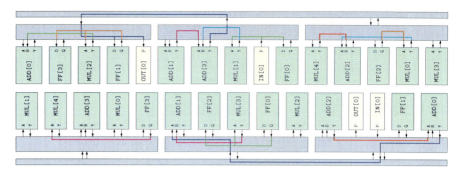

Fig. 12.4 Exemplary interconnect for the digital filters shown in Fig. 12.6 using two parallel interconnect trees (*top* and *bottom*). The routed signal paths show the post-silicon configuration for the biquad-df1 filter. In the pre-silicon phase, the interconnect was optimized with the other three topologies. Note that the cells in the *bottom half* are the same as the cells in the *top half*, but in a different order, because they are mapped to different leaves in the second interconnect tree

12.4.3 Analysis of the Tree Topology

In this section the tree topology as described in the previous section is evaluated. A set of six requirements is presented and the fitness of the tree topology to meet this requirements is analyzed.

1. Requirement: Allow random connections of the cells up to a certain degree.

   The set of netlists that can be implemented by a given interconnect tree is only limited by the number of connections between the switches and their parent switches and the tree layout (number of levels and degree of switches). An interconnect with only one big root switch is equivalent to a full-MUX interconnect that can implement any netlist. This might be useful for connection types with only a small number of input or output ports.

2. Requirement: Allow optimization of the interconnect for recurring pattern and similarities in the example netlists.

   The interconnect can be optimized towards the similarities in the example netlists by choosing cell to tree leaf mappings in a way that minimizes the interconnect utilization of the example netlists.

3. Requirement: Can be characterized using a relative simple and regular data structure. The existence of such a representation allows for easy manipulation and investigation of the interconnect topology.

   The whole interconnect can be described using only two simple data sets: Firstly the mapping of each cell to one leaf in each tree and secondly for each switch the number of connections to and from its parent switch. The first data set can be charaterized as a per-tree permutation and can be manipulated and optimized easily by exchanging the assignments of two cells in one tree. The second data set is a list of integers where greater value implies more flexibility in the post-silicon phase, but also more chip resources.

4. Requirement: Prohibits over-optimization towards the example netlists that would prevent the interconnect to work with netlists that have similarities with, but are not identical to any, example netlist.

As whole cells (instead of individual ports) are mapped to tree leaves, the optimization potential towards the individual datapaths is limited. There will always be nets that cannot be routed by only using the lowest layer of the interconnect. Thus, smart grouping of cells can be used to optimize, up to a certain degree, the interconnect to the requirements of the example netlists. On the other hand, the interconnect will not be limited to the example netlists.

5. Requirement: Allows for easy oversizing of the interconnect resources to broaden the spectrum of implementable netlists.

Oversizing is done by increasing the number of connections between switches. Netlists that are similar but not part of the set of example netlists might have nets which result in a high routing length. With oversizing, extra routing resources help to improve these cases.

6. Requirement: Easy to implement with currently available logic synthesis tools.

The interconnect topology provides only unidirectional links. This allows for an implementation using MUXes built from standard cells, as generated by ASIC synthesis tools. For the interconnect in most up-to-date FPGAs, unidirectional links are also reconsidered [9].

An additional problem arises from potential combinational loops within the interconnect circuitry. This is eliminated by forbidding to route a signal back to its originating switch. On the other hand it is still possible to create loops through combinational cells connected to the interconnect. This issue must be taken care of by disabling timing arcs through these cells and applying maximum delay constraints [2].

In summary, the chosen tree topology seams to be well suited for heterogeneous coarse-grain reconfigurable architectures.

12.5 Interconnect Synthesis

A tool called InterSynth, which automatically generates the interconnect for the reconfigurable module, was implemented. It uses a set of example netlists (each representing an actual application, compare Sect. 12.2.1) with instances of cell types and connections among them. These are used to optimize the interconnect to provide cells and connectivity, suitable for implementing any of these netlists. The output is a synthesizable Verilog file that instantiates the cells and describes the reconfigurable interconnect.

In the pre-silicon phase, the algorithm first builds the interconnect topology with the given number of parallel trees, height of the trees and order of each level. The total number of leaves is given by the number of cells required by the example netlists. Then the cells are assigned to leaves in the interconnect trees (*cell-to-leaf-mapping*) and the required number of connections for each switch to and from its parent switch are determined so that the connections of all example netlists can be routed. In that course the algorithm also implements all example netlists. This means that for each netlist, each node is mapped to a cell (*node-to-cell-mapping*) and each net is routed via one of the interconnect trees.

12.5.1 Optimization Algorithm

During the interconnect optimization, the cell-to-leaf-mappings are permuted, so that a smaller number of connections to and from the parent switches (and therefore hardware resources) is required to still implement all example netlists. This is preformed using an iterative algorithm, a single iteration of which is shown in Fig. 12.5. It operates on the state S, which contains all node-to-cell-mappings for all netlists and all cell-to-leaf-mappings for all interconnect trees.

The optimization is based on the Kernighan-Lin algorithm [8], which is an heuristic procedure for solving partitioning problems by permuting the domain mappings of entities. In InterSynth it is used (in a slightly modified manner) to permute the node-to-cell- and cell-to-leaf-mappings in the state S. The function KERNLINOPTIMIZE in Fig. 12.5 implements the Kernighan-Lin algorithm.

For the first iteration of the algorithm a start state S with random mappings is used. For all further iterations the result of the previous iteration is used as a starting point. Experiments have shown that less than six iterations are usually enough for InterSynth to reach a stable state, whereas further iterations don't significantly improve the algorithms result.

The algorithm is controlled through the use of flags that enable or disable certain parts of the algorithm. Note that the KERNLINOPTIMIZE function is using different optimization goals in different parts of the algorithm. For example the term *best candidate pair* in KERNLINOPTIMIZE is using a different definition of *best* depending on the calling block. The flag `mode_align_netlists` enables a block that "aligns" the netlists so that similar subcircuits are mapped to the same set of cells. In this block the optimization goal for KERNLINOPTIMIZE is to minimize the number of unique pairs of connected cell ports over all netlists. The flag `mode_-swap_cell_mappings` enables a block that permutes the cell-to-leaf-mappings for the individual interconnect trees and the flag `mode_swap_node_mappings` permutes the node-to-cell-mappings. In both blocks the optimization goal is to minimize the sum of the total routing lengths for all netlists in the top i levels of the interconnect trees. Therefore the first iteration of the i-loop only tries to reduce the utilization of the root switch and further iterations of the i-loop refine this first solution with respect to the other switching levels in a top-down manner.

For the *pre-silicon* procedure the algorithm is used with the flag `mode_-align_netlists` enabled in the first iteration. Thus the actual algorithm is using

12 Design Methodology and Interconnect Synthesis for Reconfigurable Architectures

Fig. 12.5 InterSynth algroithm

$S \leftarrow$ initial state

function KERNLINOPTIMIZE(S, P, T)
    $j \leftarrow 1$
    $S_0 \leftarrow S$
    while P contains compatible pairs **do**
        $S_j \leftarrow S_{j-1}$
        $(p_1, p_2) \leftarrow$ best candidate pair from P
        Swap T mapping of p_1 and p_2 in S_j
        Remove p_1 and p_2 from P
        $j \leftarrow j + 1$
    end while
    $S \leftarrow$ best candidate from $S_0 \ldots S_{j-1}$
end function

if mode_align_netlists **then**
    repeat
        $S_{\text{old}} \leftarrow S$
        for all $N =$ example netlist **do**
            $P \leftarrow$ set of all nodes in N
            KERNLINOPTIMIZE($S, P, node\_to\_cell$)
        end for
    until $S_{\text{old}} = S$
end if

for $i = 1 \rightarrow$ max. interconnect levels **do**
    if mode_swap_cell_mappings **then**
        for all $I =$ interconnects with min. i levels **do**
            $P \leftarrow$ set of all leaves in I
            KERNLINOPTIMIZE($S, P, cell\_to\_leaf$)
        end for
    end if
    if mode_swap_node_mappings **then**
        for all $N =$ example netlist **do**
            $P \leftarrow$ set of all nodes in N
            KERNLINOPTIMIZE($S, P, node\_to\_cell$)
        end for
    end if
end for

aligned netlists as a starting point. The flag mode_swap_cell_mappings is set for all iterations and mode_swap_node_mappings is only set for the second half of iterations. Thus the algorithm first tries to find a good solution without modifying the aligned netlists and after that uses this solution as a starting point for an optimization run with all degrees of freedom. After this the number of required connections for each switch to and from its parent switch is calculated by using the maximum number of these connections used for each switch in the routing results generated by the algorithm. InterSynth also provides configuration options for oversizing.

In *post-silicon* runs the flag mode_align_netlists is never activated, as there is only one netlist in post-silicon runs. The flag mode_swap_cell_-mappings is also never set during the post-silicon procedure, as the cell-to-leaf-mappings cannot be changed once the chip has been manufactured. The flag `mode_swap_node_mappings` is set in all iterations of the post-silicon procedure. As information about the available routing resources is available during the post-silicon procedure this information is used by the post-silicon routing algorithm. Thus the post-silicon routing algorithm does not optimize for the shortest path but for least congestion.

12.5.2 Implementation Details

The actual implementation of InterSynth is using performance optimizations. For example, instead of copying S to $S_0, \ldots, S_{j-1}$, a journal of the swaps is maintained that can be rolled back to the best solution. When the number of utilized nodes of a certain type varies between the netlists, additional "dummy nodes" are added by InterSynth to level the number of used nodes across all netlists. This is necessary as InterSynth can only permute the existing cell-to-leaf-mappings. That means there must be mappings for all leafs in all trees in the initial state in order to make all possible mappings accessible to the optimization algorithm. The cell type descriptions used by InterSynth provide a flag to mark a cell input as possible feedback input. An input that does not have this flag set cannot be connected directly to an output from the same cell. For most cell types such connections would never be part of a valid netlist. The Verilog HDL code generated by InterSynth can be used as-is in the final ASIC design as InterSynth can be configured to not only include the cell instantiations and interconnect logic but also additional support code in the HDL output, such as connections of cell ports to ports of the generated module (for input and output purposes or distributing global signals such as clock and reset). It is also possible to embed configuration data for reconfigurable cells (ALUs, etc.) within the InterSynth config bitstream. Inputs and outputs of the whole reconfigurable modules are handled as special cell types and therefore are not explicitly drawn in Figs. 12.2 and 12.3. The automatically generated interconnect shown in Fig. 12.4 has only one input and one output labeled `IN[0]` and `OUT[0]`.

12.6 Evaluation of InterSynth

Two different application classes were used to evaluate InterSynth: digital filters (see Sect. 12.6.1) and logic functions (see Sect. 12.6.2). For both an identical interconnect configuration was used, which has two parallel trees of height three (although with different connection types). The switches in the bottom two layers have a degree of four and the top level (root) switch connects all switches of the second layer. In order to create more flexible interconnects, an oversizing rule for one additional connection to each switch to and from its parent switch was used.

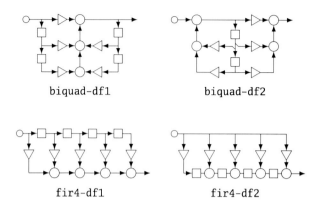

Fig. 12.6 Filter topologies used as test netlists. *Circles* represent adders, *squares* represent delays and *triangles* represent configurable constant factor multipliers

12.6.1 Filter Networks

Two instances of the four different digital filter topologies as shown in Fig. 12.6 were concatenated in all 16 possible combinations to build netlists of higher-order filters. The cell types employed are (word-wide) adders, multipliers and flip-flops. The Verilog code for this test cases can be found in Listings 12.1.

Test 1 From the pool of 16 netlists a random sample of n was selected and used for the pre-silicon phase to optimize the interconnect. Then the post-silicon phase was attempted with each of the 16 netlists. This test was performed 1,000 times each for $n \in \{1, \ldots, 6\}$. The percentage of failed post-silicon runs per post-silicon netlist and number of pre-silicon netlists is shown in the center part of Table 12.1. It shows that increasing the number of example netlists n in the pre-silicon phase results in less failed attempts in the post-silicon phase. The average resource usage of the generated interconnect is expressed with two figures: the number of bits of the configuration data and the number of 2-to-1 MUXes (MUX2) required to build the interconnect. Both numbers are normalized to the total number of cell ports. The bottom part of Table 12.1 gives their mean for $n \in \{1, \ldots, 6\}$. For the case of $n = 1$ pre-silicon netlist, the average number of MUX2 and configuration bits is shown in the right part of the table for every pre-silicon netlist.

The test also shows that post-silicon implementation of the topology fir4-df2. fir4-df2 fails in a significant fraction of the generated interconnects, especially for $n \leq 2$. This can be explained by the differences in the fir4-df2 topology compared to the other three topologies in Fig. 12.6: All multipliers in fir4-df2 are driven directly from the input (which therefore has a fanout of five) and all delay outputs are connected to adder inputs while in the other topologies delay outputs are connected to delay or multiplier inputs. It is worth mentioning that fir4-df2.fir4-df2

Table 12.1 Filter network post-silicon errors and resource usage vs. number of pre-silicon netlists

Topology	Number of pre-silicon netlists						Single	
	1	2	3	4	5	6	mux2	bits
biquad-df1.biquad-df1	2.4%	0.1%	0.0%	0.0%	0.0%	0.0%	5.74	4.07
biquad-df1.biquad-df2	1.8%	0.0%	0.0%	0.0%	0.0%	0.0%	5.80	4.08
biquad-df1.fir4-df1	2.9%	0.0%	0.0%	0.0%	0.0%	0.0%	5.75	4.07
biquad-df1.fir4-df2	2.5%	0.1%	0.0%	0.0%	0.0%	0.0%	5.85	4.11
biquad-df2.biquad-df1	2.2%	0.0%	0.0%	0.0%	0.0%	0.0%	5.81	4.11
biquad-df2.biquad-df2	0.8%	0.0%	0.0%	0.0%	0.0%	0.0%	5.73	4.06
biquad-df2.fir4-df1	3.5%	0.1%	0.0%	0.0%	0.0%	0.0%	5.77	4.08
biquad-df2.fir4-df2	3.0%	0.1%	0.0%	0.0%	0.0%	0.0%	5.80	4.11
fir4-df1.biquad-df1	13.4%	0.5%	0.0%	0.0%	0.0%	0.0%	5.88	4.13
fir4-df1.biquad-df2	3.4%	0.0%	0.0%	0.0%	0.0%	0.0%	5.77	4.06
fir4-df1.fir4-df1	14.7%	0.4%	0.0%	0.0%	0.0%	0.0%	5.86	4.11
fir4-df1.fir4-df2	4.1%	0.0%	0.0%	0.0%	0.0%	0.0%	5.85	4.12
fir4-df2.biquad-df1	6.0%	0.0%	0.0%	0.1%	0.0%	0.0%	5.86	4.13
fir4-df2.biquad-df2	1.7%	0.1%	0.0%	0.0%	0.0%	0.0%	5.77	4.08
fir4-df2.fir4-df1	4.4%	0.2%	0.0%	0.1%	0.0%	0.0%	5.83	4.10
fir4-df2.fir4-df2	34.5%	6.2%	1.6%	0.5%	0.2%	0.3%	5.86	4.13
avg. mux2/port	5.82	7.15	8.24	9.40	10.31	11.04		
avg. bits/port	4.10	4.61	5.01	5.43	5.76	6.01		

does not require more routing resources than the other topologies (see right part of Table 12.1). It only requires a different interconnect because it is composed of different patterns. Thus an interconnect that can implement `fir4-df2.fir4-df2` as well as the other 15 topologies needs more resources than one that can only implement the 15 others.

Test 2 The resource usage of the pre-silicon results where compared to the resource usage of an interconnect with a random, i.e., unoptimized cell-to-leaf-mappings (`mode_swap_cell_mappings` disabled in all iterations of the algorithm). The difference in the resources needed for these two cases is an indicator of the optimization potential utilized by InterSynth to optimize the interconnect for the application domain described by the example netlists. When $n = 4$ pre-silicon netlists are used and *no* additional routing resources are added, an average number of 3.0 (stddev 1.1) word-wide MUX2 per cell port are required to implement the filter example. When the InterSynth cell to leaf mapping is replaced with a random mapping and InterSynth is only used for the node-to-cell-mappings, this number increases to 7.2 (stddev 0.6). This shows that InterSynth can drastically optimize interconnects for scenarios like this one with a relatively large number of cell types compared to the number of cells.

Test 3 The number of parallel interconnect trees was varied from one to four and the degree of the switches was varied from two (binary tree) to six. For each

12 Design Methodology and Interconnect Synthesis for Reconfigurable Architectures

Table 12.2 Number of trees and degree of switches vs. interconnect resource usage and post-silicon errors

		Degree of interconnect trees					Degree of interconnect trees				
		2	3	4	5	6	2	3	4	5	6
# of trees	1	10.74	8.39	8.42	7.21	8.34	116	82	81	43	32
	2	8.51	7.86	8.27	8.31	9.14	46	34	5	6	2
	3	8.48	8.57	9.75	10.39	11.80	0	0	0	0	0
	4	9.57	10.21	12.23	13.22	15.22	0	0	0	0	0
		MUX2/port					**Post-silicon errors/1k**				

of these interconnect configurations, two random pre-silicon netlists where used for optimization. The average resource utilization (number of MUX2) for each configuration is given in the left part of Table 12.2.

Each optimized interconnect was used for 1,000 post-silicon netlists. The number of errors (i.e. the post-silicon netlist could not be routed within the interconnect) is shown in the right part of Table 12.2. For a single interconnect tree, even with a high degree of switches, a large number of post-silicon errors are present. Two parallel trees and a degree of four and above result in an acceptable number of post-silicon errors. Therefore two parallel trees with switches of degree four are a trade-off with resource utilization. More parallel trees result in a large increase of resource utilization and might also result in a wiring congestion on chip in larger scenarios.

12.6.2 Logic Networks

Random logic functions with six inputs and one output were generated and ABC [1] was used to convert these logic functions to netlists of inverters, two-input AND gates and two-input XOR gates. Of course such a problem would be better solved by rather using lookup tables than configurable interconnects and basic logic gates, but this is a simple method for generating a virtually unlimited pool of "similar" large netlists. For this test InterSynth was configured with oversizing rules to add 10% plus 5 cells of each kind to compensate for the variation in the cell usage in the generated netlists.

Test 1 For the pre-silicon phase, four random example netlists were used to optimize the interconnect. The results from this pre-silicon phase were then tested using 1,000 other random netlists (limited by the number of available cells) for the post-silicon phase. This was performed 50 times. The post-silicon run failed in only 0.05% of these 50,000 tests.

Test 2 An average number of 16.8 (stddev 0.8) MUX2 per cell port are required to implement this testcase (with four pre-silicon netlists) regardless of the question whether the cell-to-leaf-mapping was optimized or not (i.e., `mode_swap_cell_mappings` was enable or disabled). This shows that while it is possible to use InterSynth for large homogeneous networks like this test case, it doesn't have an advantage over distributing the cells regularly.

12.7 Yosys

In order to provide a convinient way for design entry, a feature-rich HDL synthesis tool with the name Yosys[1] was implemented. Yosys is a generic Verilog synthesis tool[2] that can be used in a wide variety of application domains [14].

The Verilog code in Listings 12.1 was used to create the InterSynth netlists for the filters used in the evaluation presented in the last section. Additional input files for Yosys include a small synthesis script and an additional Verilog file that describe how Yosys should map the RTL constructs to the coarse grain cell library.

Besides simple HDL synthesis Yosys can be used for a wide range of advanced analyzes and circuit transformations. It can extract FSMs and perform various operations on extracted FSMs, such as recoding and moving additional function from logic networks into the FSM. In coarse-grain environments this can be used to move control logic into a generic FSM cell, e.g. TR-FSM [6].

Yosys also supports technology mapping by finding subcircuit isomorphism, allowing coarse-grain cells to implement richer logic function than the RTL cells used by Yosys internally. Yosys also has limited support for frequent subcircuit mining, easing the identification of possible coarse-grain cell types during the pre-silicon design phase.

12.8 Conclusion

A design methodology for application-domain specific heterogeneous coarse-grain reconfigurable logic architectures is presented. One or multiple such resulting reconfigurable modules are integrated into an SoC to off-load its CPU. This results in a large reduction of power consumption. Contrary to FPGAs, a coarse-grain and heterogeneous architecture is used, which allows further reduction in power and area.

In the pre-silicon phase, the application class for the reconfigurable module is defined and specified by several example applications. These are synthesized and analyzed to extract common logic structures as coarse-grain cells (including

---

[1] A left-recursive acronym for "Yosys Open Synthesis Suite".

[2] VHDL support is in development as of this writing.

12 Design Methodology and Interconnect Synthesis for Reconfigurable Architectures

Listing 12.1 Verilog code for generating the filter netlists

```verilog
module filter(input clk, input [31:0] in, output reg [31:0] out);
    parameter type = 0;
    parameter k1 = 1, k2 = 2, k3 = 3, k4 = 4, k5 = 5;
    reg [31:0] next_tmp [3:0], tmp [3:0];
    integer i;
    always @*
        case (type)
            0: begin // biquad-df1
                next_tmp[0] <= in;  next_tmp[1] <= tmp[0];
                next_tmp[2] <= out; next_tmp[3] <= tmp[2];
                out <= k1*in + k2*tmp[0] + k3*tmp[1] +
                              k4*tmp[2] + k5*tmp[3];
            end
            1: begin // biquad-df2
                next_tmp[0] <= in + k1*tmp[0] + k2*tmp[1];
                next_tmp[1] <= tmp[0];
                out <= k3*in + k4*tmp[0] + k5*tmp[1];
            end
            2: begin // fir4-df1
                for (i = 0; i < 4; i = i+1)
                    next_tmp[i] <= i > 0 ? tmp[i-1] : in;
                out <= k1*in + k2*tmp[0] + k3*tmp[1] +
                              k4*tmp[2] + k5*tmp[3];
            end
            3: begin // fir4-df2
                next_tmp[0] <= in*k1;
                next_tmp[1] <= in*k2 + tmp[0];
                next_tmp[2] <= in*k3 + tmp[1];
                next_tmp[3] <= in*k4 + tmp[2];
                out          <= in*k5 + tmp[3];
            end
        endcase
    always @(posedge clk)
        for (i = 0; i < 4; i = i+1)
            tmp[i] <= next_tmp[i];
endmodule

module filter2(input clk, input [31:0] in, output reg [31:0] out);
    parameter type = 0;
    wire [31:0] tmp;
    filter #( .type(type % 4) )
        F1 (.clk(clk), .in(in),  .out(tmp) );
    filter #( .type(type / 4) )
        F2 (.clk(clk), .in(tmp), .out(out) );
endmodule

module top;
    genvar i;
    generate for (i = 0; i < 16; i = i+1) begin:list
        filter2 #( .type(i) ) F ();
    end endgenerate
endmodule
```

reconfigurable FSMs) and to build a cell library. The example application circuits are transformed to only instantiate such coarse-grain cells.

The major part of this work presents an algorithm to merge these example application netlists to a single reconfigurable module. It optimizes a tree structured interconnect and the selection of coarse-grain cells which are able to implement all example applications. Spending additional hardware resources even allows to implement yet-unknown applications with the resulting silicon.

The evaluation of the algorithm was performed using digital filter topologies. With only two example netlists and slight oversizing in the pre-silicon phase, nearly all other example netlists could be realized in the post-silicon phase. Additionally, a large optimization potential to keep the hardware resources limited was demonstrated.

We propose improvements to InterSynth in the following areas: The routing algorithm for the pre- and post-silicon phases can be improved, for example to support routing of a single net in multiple trees.

The over-all optimization procedure can also be improved: As depicted in the right part of Table 12.1, the hardware resources (MUX2) of the interconnect increase when more pre-silicon netlists are used, even when all example netlists can be routed. A consolidation step after the pre-silicon procedure would help reduce the hardware resources in this cases.

InterSynth is a generic tool for creating interconnects using the procedure described in this work. It is implemented in C++ and released as an Open Source project at http://www.clifford.at/intersynth/. The scripts used to run the experiments in Sect. 12.6 are included.

Yosys is a generic versatile tool for digital circuit synthesis. Besides its other uses, it can be used as Verilog-frontend for InterSynth as well as for circuit analysis in the pre-silicon and design phase. Is is also released as an Open Source project at http://github.com/cliffordwolf/yosys.

Acknowledgements This work has been supported (in part) by the Austrian COMET K-project ECV under contract no. 815105.

References

1. Berkeley Logic Synthesis and Verification Group: ABC: a system for sequential synthesis and verification. [Mercurial checkout 13 February 2012], http://www.eecs.berkeley.edu/~alanmi/abc/
2. Bhatnagar, H.: Advanced ASIC Chip Synthesis Using Synopsys Design Compiler, Physical Compiler, and PrimeTime. Kluwer, Boston (2002)
3. Ferreira, R., Vendramini, J.G., Mucida, L., Pereira, M.M., Carro, L.: An FPGA-based heterogeneous coarse-grained dynamically reconfigurable architecture. In: Proceedings of the 14th International Conference on Compilers, Architectures and Synthesis for Embedded Systems (CASES), Taipei, Oct 2011, pp. 195–204
4. Glaser, J., Haase, J., Damm, M., Grimm, C.: Investigating power-reduction for a reconfigurable sensor interface. In: Proceedings of Austrochip 2009, Graz, 7 Oct 2009

5. Glaser, J., Haase, J., Grimm, C.: Designing a reconfigurable architecture for ultra-low power wireless sensors. In: Ghassemlooy, Z., Ng, W.P. (eds.) Proceedings of the Seventh IEEE, IET International Symposium on Communication Systems, Networks and Digital Signal Processing (CSNDSP), Northumbria University, Newcastle upon Tyne, 21–23 July 2010, pp. 343–347

6. Glaser, J., Damm, M., Haase, J., Grimm, C.: TR-FSM: transition-based reconfigurable finite state machine. ACM Trans. Reconfigurable Technol. Syst. (TRETS) **4**(3), 23:1–23:14 (2011)

7. Glaser, J., Gravogl, K., Haase, J., Grimm, C.: A reconfigurable architecture for ultra-low power wireless sensors. Mediterr. J. Electron. Commun. (MEDJEC) **7**(3), 255–266 (2011)

8. Kernighan, B.W., Lin, S.: An efficient heuristic procedure for partitioning graphs. Bell Syst. Tech. J. **49**(1), 291–307 (1970)

9. Lemieux, G., Lee, E., Tom, M., Yu, A.: Directional and single-driver wires in FPGA interconnect. In: Proceedings of the IEEE International Conference on Field-Programmable Technology, Brisbane, pp. 41–48. IEEE, (2004)

10. Marrakchi, Z., Mrabet, H., Farooq, U., Mehrez, H.: FPGA interconnect topologies exploration. Int. J. Reconfigurable Comput. **2009**, 1–13 (2009)

11. Mehta, G., Stander, J., Lucas, J., Hoare, R.R., Hunsaker, B., Jones, A.K.: A low-energy reconfigurable fabric for the SuperCISC architecture. J. Low Power Electron. **2**(2), 148–164 (2006)

12. Ou, J., Muhammad, F., Haase, J., Grimm, C.: A technique for the identification of reconfigurable resources of flexible communication systems. In: NASA/ESA Conference on Adaptive Hardware and Systems (AHS), San Diego, June 2011, pp. 256–263

13. Wolf, C., Glaser, J., Schupfer, F., Haase, J., Grimm, C.: Example-driven interconnect synthesis for heterogeneous coarse-grain reconfigurable logic. In: Forum on Specification and Design Languages (FDL), Vienna, 18–20 Sept 2012, pp. 194–201

14. Wolf, C., Glaser, J.: Yosys – A Free Verilog Synthesis Suite. Submitted to: Proceedings of the 21st Austrian Workshop on Microelectronics (Austrochip), Linz, Austria, 10 Oct 2013

15. ul Abdin, Z., Svensson, B.: Evolution in architectures and programming methodologies of coarse-grained reconfigurable computing. Microprocess. Microsyst. **33**(3), 161–178 (2009)

Printed by Publishers' Graphics LLC
LMO131003.15.17.14